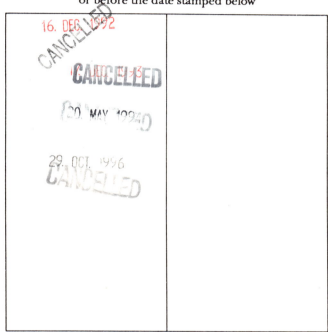

Manual on
Experimental Methods
for
Mechanical Testing
of Composites

The Manual on Experimental Methods for Mechanical Testing of Composites, published by the Society for Experimental Mechanics, Inc., is a project of the Composite Materials Division of SEM, Inc.

Manual on
Experimental Methods for Mechanical Testing of Composites

Edited by

Richard L. Pendleton
South Dakota School of Mines and Technology
Rapid City, South Dakota, U.S.A.

Mark E. Tuttle
University of Washington
Seattle, Washington, U.S.A.

Elsevier Applied Science Publishers
London, U.K.

Society for Experimental Mechanics, Inc.
Bethel, Connecticut, U.S.A.

Kenneth A. Galione, Publisher

Sole Distributor Outside the USA and Canada
ELSEVIER SCIENCE PUBLISHERS LTD.
Crown House, Linton Road, Barking
Essex IG11 8JU, U.K.

Sole Distributor in the USA and Canada
Society for Experimental Mechanics, Inc. 1989
7 School Street, Bethel
CT 06801, U.S.A.

The selection and presentation of material and the opinions expressed in this publication are the sole responsibility of the authors concerned.

ISBN 1-85166-3754
ISBN 0-912053-23-2

British Library Cataloguing in Publication Data Manual
on experimental methods for mechanical testing of composites.
  1. Composite materials. Testing
  I. Pendleton, Richard L.   II. Tuttle, Mark E.
  620.1'18

Table of Contents

Manual on Experimental Methods for Mechanical Testing of Composites

Introduction

by Richard L. Pendleton

Contents and Format of This Manual

The goal of this manual is to provide a series of basic tutorial chapters which describe practical details involved in applying various experimental techniques to composite materials and composite structures. These tutorial chapters are not intended to extend the state-of-the-art but to allow a well-trained, experienced individual to enter the world of experimental composite materials without inventing numerous wheels that have already been invented.

Overview

During the past 100 years the science and art of materials engineering and mechanical design have been progressively developed by the ingenuity and perseverance of a multitude of dedicated people. Many of us have applied ourselves diligently to learn and develop complex principles, both analytical and experimental, for predicting the behavior and failure of a wide variety of materials which are essentially isotropic. Work of the past generation has led to an explosive materials revolution which allows products and equipment to be engineered and operated at strength-to-weight ratios unheard of in the past. Materials can be contoured and precisely tailored to meet specific strength or stiffness requirements. Furthermore, they can be constructed directionally, i.e., a composite laminate can be constructed to have virtually optimal strength not only in the direction of primary loads or stresses but in orthogonal directions as well. High-performance composites are often highly orthotropic (or transversely isotropic) and inhomogenous materials. Since most practical engineering experience is based upon the familiar behavior of isotropic materials, composites can exhibit surprising and unusual behavior which may lead to erroneous interpretation of experimental results. Consequently, **caution must be exercised when studying the behavior of composites using an experimental method previously developed for use with conventional isotropic materials.**

The continued research and development in the materials sciences incurs enormous difficulties for both the practicing industrial engineer or teacher in our colleges and universities because many, perhaps most, of our previous analysis and testing techniques are either totally without benefit or need to be significantly modified for use with composite materials. The Composite Materials Division of the Society for Experimental Mechanics (SEM) is presenting this manual on "Experimental Methods for Mechanical Testing of Composites" to provide assistance to those who are entering the field of composite materials and require instruction in these experimental methods. Note that the manual is not intended to be a composite materials textbook, nor are the 'standard' specimen configurations or test procedures used in industry discussed. The manual is devoted entirely to fundmental experimental methods or tools; how these methods might be used to verify or supplement a theory of some sort (e.g., fracture mechanics) is not discussed. The manual is sponsored by the Composite Materials Technical Division of the Society for Experimental Mechanics. Each contributor has been asked to keep analytic discussion to the minimal possible level, and to focus instead upon the practical details of the experimental technique being described. It is hoped that these articles will allow a novice to quickly grasp the underlying principles, advantages, and limitations of each experimental method, while still representing a valuable reference source for a more experienced composites engineer. The American Society of Testing Materials conscientiously and expertly develops standards for a wide variety of composite-materials testing and every researcher should be thoroughly familiar with their publications. We intend to provide here a tutorial format rather than instructions for a specific test.

Basic Description of Composite Materials

Definition

A composite material is created by combining two or more materials for the purpose of predictably enhancing certain properties. In the field of advanced-composites laminates, this combination consists of a reinforcing agent (fiber), a compatible matrix, and perhaps a filler and a binding agent.

Classification

There are three major classifications of composites[1] They are fibrous, laminar, and particulate, and are described below.

(a.) Fibrous composites are materials containing reinforcing fibers bonded to a matrix filler material. Fibers are very small in diameter and provide little or no strength or stiffness except in tension. Generally, a smaller diameter means fewer dislocations and instabilities within the fiber materials and, consequently, higher tensile strength. Many different materials are presently used as fibers, including carbon, boron, graphite and tungsten.

(b.) Laminar composites are composed of layers of materials bonded together. This category includes both sandwich and honeycomb composites as well as several types of wooden layered composites. A major area of study includes orthotropic laminates which will be discussed below.

(c.) Particulate composites consist of particles dispersed in a matrix. The types of particles can be either skeletal or flake and a wide variety of sizes, shapes and materials are available.

Orthotropic Laminate

Most composites are composed of a relatively stiff, high-strength **reinforcing material**, embedded within a relatively compliant, low-strength **matrix material**. A bewildering array of different types of composite materials is currently available to the structural engineer. One complication is that the nomenclature used to describe composite materials has not yet been wholly standardized. Composites are generally classified according to both (a) the physical size or shape of the reinforcing material, and (b) the type of matrix material. The major classifications of reinforcing and matrix materials are listed below:

Major Reinforcement Classifications:
- Roughly spherical particulates
- Flat flakes
- Whiskers, with a length (l) typically less than about 20mm
- Short (or "chopped") fibers, where 20mm < l < 150mm
- Continuous fibers, where l > 150mm
- Skeletal reinforcement

Major Matrix Classifications:
- Polymeric
- Metallic
- Ceramic

Generally Orthotropic Materials

A single layer, or lamina, of a composite laminate consists of numerous fibers embedded in, and bonded to, a matrix material. The primary strength and stiffness is derived from the fiber and is in the direction of the fiber. Clearly the material is not isotropic but neither is it entirely anisotropic due to the regular orientation of the fibers. We classify the three orthogonal directions parallel and perpendicular to the fibers as the 'principal' material directions. It is possible to develop relationships[2] based on Hooke's law between the lamina stress and strain, using assumptions of linearity, which require nine nonzero material constants (four in two dimensions). This type of material is termed 'orthotropic' or, 'generally orthotropic'. Experimental methods have been developed to determine these material constants with a reasonable amount of experimental effort.

The term "graphite/epoxy" typically refers to a combination of graphite fibers embedded in an epoxy matrix, but without further information one cannot tell whether the composite is produced using (random or oriented) graphite whiskers or short fibers, or in the case of laminated composites, whether the plies are unidirectional or woven fabrics. Although this example is based on a polymeric-matrix composite material system, similar observations can be made for metal- or ceramic-matrix composites as well.

In short, the nomenclature used to describe composites is still developing and at present is often vague and ill-defined. The reader is referred to the text by Schwarz[3] for a more detailed discussion of the various forms of composite materials currently available.

Experimental Methods

Experimental methods have traditionally been used to obtain information concerning deformations, strains, structural integrity and failure mechanisms in a solid structure or mechanism. Common experimental methods include strain gages, photoelasticity, moire, ultrasound and radiography, as well as acoustic and thermographic methods. These methods are also used with composite materials but the inclusion of such parameters as stacking sequence, orthotropy, nonhomogeniety, and inelasticity require significant adjustments, both in the methods and the interpretation of data. It is precisely because of these parameters that this manual is expected to be of substantial benefit to those academic and industrial scientists who are entering the world of composite materials. An additional factor, of fundamental importance in the field of composite laminates, is that most composite-laminate applications require that the mechanism or structure be specifically produced in a batch process. Unlike most metals which are mass produced, the properties of a composite-laminate part are unique to the quality-control methods used during the production of the part. This strongly invites the use of nondestructive testing which can be directly applied to the part being used rather than a representative sample. This requirement is presented very nicely in Carlsson and Pipes's "Experimental Characterization of Advanced Composite Materials."[4]

This manual is primarily (but not wholly) devoted to experimental methods which have been applied to laminated polymer-matrix composites. However, most of the experimental methods described can be applied to other types of composites as well. The nomenclature most frequently used to describe laminated polymeric composites is defined in the articles titled "Anisotropic Material Behavior", and "Classical Lamination Theory", which appear in the following chapter.

Acknowledgments

The editors wish to express their sincere appreciation to the many authors who have donated their time and expertise to this manual. We believe their efforts have made this manual unique among the composites literature. We also acknowledge the support and guidance provided by members of the Composite Materials Technical Division, as well as the Headquarters staff of the Society for Experimental Mechanics. Finally, we express appreciation to the University of Washington and South Dakota School of Mines and Technology for providing the publishing and travel funds required to complete the manual.

References

1. Richardson, T.L., "Composites, A Design Guide," Industrial Press Inc. (1987).
2. Bert. Charles W.. "Manual on Experimental Methods for Mechanical Testing of Composites," Society for Experimental Mechanics, Inc., 1989.
3. Schwarz, M.M., "Composite Materials Handbook," McGraw-Hill Book Company, New York, ISBN 0-07-055743-8 (1984).
4. Carlsson and Pipes, "Experimental Characterization of Advanced Composite Materials," Prentice-Hall Inc. (1987).

Section IIA

Anisotropic-Material Behavior

by Charles W. Bert

Introduction

By their very nature, composites are nonhomogeneous bodies, consisting of reinforcements (such as the fibers or particles), the surrounding matrix (a polymer, metal, or ceramic), and a coupling agent which bonds them together. However, the scale of the reinforcement is usually quite small (a few mils in nominal diameter), so that, from the standpoint of engineering design as well as experimental mechanics, it is convenient to consider an individual layer of composite to be macroscopically homogeneous. This is analogous to the usual consideration of metallic structural materials, which are actually crystalline aggregates, as homogeneous materials. However, in the case of composites, there is one fundamental difference: the macroscopic behavior is usually directionally dependent. (The exception is particulate composites, which will not be mentioned further; they can often be considered to be macroscopically isotropic). A material having directionally dependent behavior is said to be *anisotropic*, i.e., nonisotropic.

Since composites, with the possible exception of metal-matrix composites, exhibit nearly linear behavior almost to failure, the present discussion is limited to linear-elastic materials, in which stress is linearly proportional to strain. In the most general case, the three-dimensional Hooke's Law can be written as a set of linear equations in which the six strains (three normal and three shear) can each be written as the sum of six terms linearly proportional to the six stresses (three normal and three shear). This can be expressed as a six-by-six array of algebraic equations involving 36 material constants. However, it can be shown by strain-energy considerations that the array possesses symmetry so that only 21 of the material constants are independent in the most complicated case.

Fortunately, the type of anisotropic behavior exhibited by composites is not nearly as complicated as that of single crystals. In fact, the nature of the geometry of unidirectionally reinforced composites, in which all of the fibers are nominally parallel to each other, is what is known as *orthotropic*; i.e., there are three mutually perpendicular planes of elastic symmetry. Then, the number of independent elastic constants is reduced from 21 to only nine.

The appropriate type of coordinate system depends upon the geometry of the reinforcing array. However, in the most common case, it is convenient to use rectangular Cartesian coordinates, denoted by 1, 2, and 3. These coordinates are utilized in the following subsections.

Orthotropic Material

In elementary solid mechanics, it is customary to denote normal stresses by the symbol σ and shear stresses by the symbol τ. Similarly, normal strains are denoted by the symbol ε and shear strains by γ. Normal stress is subscripted by a symbol denoting the direction of the axis which is perpendicular to the plane on which it acts. For instance, σ_1 denotes the normal stress acting in direction 1. Shear stresses are traditionally subscripted by two subscripts: the first one denotes the direction of the normal to the plane on which it acts and the second one denotes the direction in which the shear stress acts. Thus, there are two shear stresses which act on a plane perpendicular to the 1 axis: τ_{12} and τ_{13}, which act in the 2 and 3 directions, respectively. For equilibrium, it is necessary that $\tau_{21} = \tau_{12}$, $\tau_{32} = \tau_{23}$, and $\tau_{13} = \tau_{31}$. Thus, as is well known, there are only three independent shear stresses. A cube depicting the six stress components is shown in Fig. 1.

Analogous to the stresses, the normal strains are traditionally denoted by one subscript and the shear strains by two. Also, $\gamma_{21} = \gamma_{12}$, $\gamma_{32} = \gamma_{23}$, and $\gamma_{13} = \gamma_{31}$, so that there are only three independent shear strains.

In addition to the strains induced by mechanical stress, there may be strains induced by changes in temperature and in moisture concentration, denoted by ΔT and ΔC, respectively.

For the case of an orthotropic material subjected to mechanical, thermal and hygrothermal actions, the thermoelastic constitutive relation can be written

$$\begin{Bmatrix} \varepsilon_1 \\ \varepsilon_2 \\ \varepsilon_3 \\ \gamma_{23} \\ \gamma_{31} \\ \gamma_{12} \end{Bmatrix} = \begin{bmatrix} 1/E_1 & \nu_{21}/E_2 & \nu_{31}/E_3 & 0 & 0 & 0 \\ \nu_{12}/E_1 & 1/E_2 & \nu_{32}/E_3 & 0 & 0 & 0 \\ \nu_{13}/E_1 & \nu_{23}/E_2 & 1/E_3 & 0 & 0 & 0 \\ 0 & 0 & 0 & 1/G_{23} & 0 & 0 \\ 0 & 0 & 0 & 0 & 1/G_{31} & 0 \\ 0 & 0 & 0 & 0 & 0 & 1/G_{12} \end{bmatrix} \begin{Bmatrix} \sigma_1 \\ \sigma_2 \\ \sigma_3 \\ \tau_{23} \\ \tau_{31} \\ \tau_{12} \end{Bmatrix} \tag{1}$$

$$+ \begin{Bmatrix} \alpha_1 \\ \alpha_2 \\ \alpha_3 \\ 0 \\ 0 \\ 0 \end{Bmatrix} \Delta T + \begin{Bmatrix} \beta_1 \\ \beta_2 \\ \beta_3 \\ 0 \\ 0 \\ 0 \end{Bmatrix} \Delta C$$

where

E_1, E_2, E_3	$\equiv$	elastic moduli in the 1,2,3 directions
G_{23}, G_{31}, G_{12}	$\equiv$	shear moduli
$\nu_{12}, \nu_{21}, \nu_{23}, \nu_{32}, \nu_{31}, \nu_{13}$	$\equiv$	Poisson's ratios
$\alpha_1, \alpha_2, \alpha_3$	$\equiv$	thermal expansion coefficients
$\beta_1, \beta_2, \beta_3$	$\equiv$	moisture expansion coefficients

It is noted that there are 12 elastic coefficients (three E's, three G's, and six ν's) listed. However, only nine of them are independent, since the array of elastic coefficients must be symmetric, as mentioned previously. Thus, the following three so-called reciprocal relations must hold:

$$\nu_{21}/E_2 = \nu_{12}/E_1$$
$$\nu_{32}/E_3 = \nu_{23}/E_2 \tag{2}$$
$$\nu_{31}/E_3 = \nu_{13}/E_1$$

Then there are only nine *independent* elastic constants. Also, it is noted that relative to this coordinate system, in which 1, 2, and 3 are material-symmetry directions, there are many zeros in eq (1).

It is unwieldy to use the above notation for composites, so a contracted notation system is widely used. In this system,

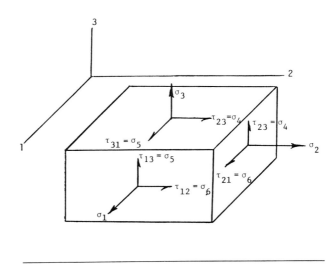

Fig. 1—Three-dimensional stresses in rectangular coordinates

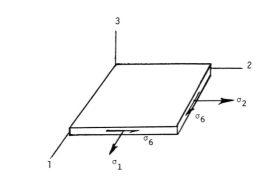

Fig. 2—Plane-stress element

$$\varepsilon_1, \varepsilon_2, \varepsilon_3 \rightarrow \varepsilon_1, \varepsilon_2, \varepsilon_3$$
$$\sigma_1, \sigma_2, \sigma_3 \rightarrow \sigma_1, \sigma_2, \sigma_3 \tag{3}$$
$$\gamma_{23}, \gamma_{31}, \gamma_{12} \rightarrow \varepsilon_4, \varepsilon_5, \varepsilon_6$$
$$\tau_{23}, \tau_{31}, \tau_{12} \rightarrow \sigma_4, \sigma_5, \sigma_6$$

Then, eq (1) can be rewritten in compact form as

$$\{\varepsilon_{ij}\} = [S_{ij}]\{S_{ij}\}\{\sigma_j\} + \{\alpha_i\Delta T + \{\beta_i\{\Delta C \atop (i,j=1,\dots,6)} \tag{4}$$

where $\{\varepsilon_i\}$ has elements $\varepsilon_1, \dots, \varepsilon_6$; and $\{\sigma_j\}$ has elements $\sigma_1, \dots, \sigma_6$. The coefficients S_{ij} are called the compliance coefficients and can be related to the customary elastic properties by direct comparison of the corresponding terms in eqs. (4) and (1). For example:

$$S_{11} = 1/E_1$$

$$S_{12} = \nu_{21}/E_2 = S_{21} \qquad (5)$$

$$S_{66} = 1/G_{12}$$

Although eq (4) is convenient for calculating the strains from the stresses and temperature and moisture changes, it is not convenient for determining the stresses. In this case, eq (4) must be inverted, resulting in

$$\{\sigma_i\} = [C_{ij}]\left\{\{\varepsilon_j\} - \{\alpha_i\}\Delta T - \{\beta_i\}\Delta C\right\}$$
$$(i,j = 1,\ldots,6) \qquad (6)$$

where the three-dimensional stiffness coefficient matrix, $[C] = [S]^{-1}$, and $[\ \ ]^{-1}$ denotes an inverse matrix.

Plane-Stress Case

In two important instances, the general three-dimensional case, as discussed in the preceding section, can be considerably simplified: (1) at a free surface, which is the situation for surface stress analysis, as in the case of strain measurements using strain gages or photoelastic coatings, and (2) for a thin layer (sometimes called a ply or a lamina) for which the thickness direction stresses are negligible.

Both cases involve plane-stress analysis. If the plane of concern is the xy plane, the only nonzero stresses are the in-plane normal stresses σ_1 and σ_2 and the in-plane shear stress σ_6; see notations (3) and Fig. 2. It is to be emphasized that although the thickness-direction normal stress (σ_3) is zero, in general, the corresponding strain (ε_3) is not. Thus, to reduce eq (6) to the plane-stress case, one cannot just strike out rows 3, 4, and 5 and columns 3, 4, and 5 in eq (6). First, to obtain an expression for ε_3, the third row of eq (4) is used, with all stress components set equal to zero except σ_1 and σ_2. This shows that

$$\varepsilon_3 = S_{13}\sigma_1 + S_{23}\sigma_2 \qquad (7)$$

Then one can use eq (7) in the first two rows of eq (6). The final result, for all three in-plane stresses can be expressed as

$$\begin{Bmatrix} \sigma_1 \\ \sigma_2 \\ \sigma_6 \end{Bmatrix} = \begin{bmatrix} Q_{11} & Q_{12} & 0 \\ Q_{12} & Q_{22} & 0 \\ 0 & 0 & Q_{66} \end{bmatrix} \begin{Bmatrix} \varepsilon_1 - \alpha_1\Delta T - \beta_1\Delta C \\ \varepsilon_2 - \alpha_2\Delta T - \beta_2\Delta C \\ \varepsilon_6 \end{Bmatrix} \quad (8)$$

where the Q's are plane-stress reduced stiffnesses given by

$$Q_{11} = S_{22}/S_{11}S_{22} - S_{12}{}^2)$$

$$Q_{12} = -S_{12}/(S_{11}S_{22} - S_{12}{}^2)$$

$$Q_{22} = S_{11}/(S_{11}S_{22} - S_{12}{}^2) \qquad (9)$$

$$Q_{66} = 1/S_{66}$$

Using eq (5), one can express the Q's in terms of the ordinary elastic coefficients as follows:

$$Q_{11} = E_1/(1 - \nu_{12}\nu_{21})$$

$$Q_{12} = \nu_{21}E_1/(1 - \nu_{12}\nu_{21})$$

$$Q_{22} = E_2/(1 - \nu_{12}\nu_{21}) \qquad (10)$$

$$Q_{66} = G_{12}$$

It is noted that eq (8) is very useful in converting measured strains to stresses. In fact, it is the orthotropic generalization of the familiar isotropic equation

$$\begin{Bmatrix} \sigma_1 \\ \sigma_2 \\ \sigma_6 \end{Bmatrix} = \frac{E}{1-\nu^2} \begin{bmatrix} 1 & \nu & 0 \\ \nu & 1 & 0 \\ 0 & 0 & (1-\nu)/2 \end{bmatrix} \begin{Bmatrix} \varepsilon_1 - \alpha\Delta T - \beta\Delta C \\ \varepsilon_2 - \alpha\Delta T - \beta\Delta C \\ \varepsilon_6 \end{Bmatrix} \quad (11)$$

Fig. 3—Stress components in a unidirectional layer referred to material-symmetry and arbitrary directions

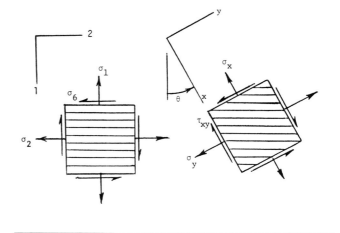

Transformation of Stresses, Strains, and Plane-Stress Reduced Stiffnesses

It is advantageous to arrange the various layers of a laminate in various directions; see Section IIB. Thus, in general, the material-symmetry axes (fiber direction and in-plane direction normal to it) do not coincide with the loading or reference directions. For this reason, it is necessary to consider stresses, strains, stiffnesses, and expansion coefficients for a coordinate system (x,y) which is rotated through an angle θ with respect to the orthotropic material-symmetry axis system (1,2); see Fig. 3.

Stresses, strains, and expansion coefficients in the x,y system can be obtained by Mohr's Circle transformations which can be written in compact form as follows:

$$\begin{Bmatrix} \sigma_x \\ \sigma_y \\ \tau_{xy} \end{Bmatrix} = [T]^{-1} \begin{Bmatrix} \sigma_1 \\ \sigma_2 \\ \sigma_6 \end{Bmatrix} \quad ; \quad \begin{Bmatrix} \varepsilon_x \\ \varepsilon_y \\ \tfrac{1}{2}\gamma_{xy} \end{Bmatrix} = [T]^{-1} \begin{Bmatrix} \varepsilon_1 \\ \varepsilon_2 \\ \tfrac{1}{2}\varepsilon_6 \end{Bmatrix}$$

(12)

$$\begin{Bmatrix} \alpha_x \\ \alpha_y \\ \tfrac{1}{2}\alpha_{xy} \end{Bmatrix} = [T]^{-1} \begin{Bmatrix} \alpha_1 \\ \alpha_2 \\ 0 \end{Bmatrix} \quad ; \quad \begin{Bmatrix} \beta_x \\ \beta_y \\ \tfrac{1}{2}\beta_{xy} \end{Bmatrix} = [T]^{-1} \begin{Bmatrix} \beta_1 \\ \beta_2 \\ 0 \end{Bmatrix}$$

where the transformation matrix is given by

$$[T] = \begin{bmatrix} m^2 & n^2 & 2mn \\ n^2 & m^2 & -2mn \\ -mn & mn & m^2 - n^2 \end{bmatrix}$$

(13)

Here, $m \equiv \cos\theta$ and $n \equiv \sin\theta$.

By a series of matrix operations (see Ref. 1 for details) involving eqs (12) and (13), one finds the new constitutive relations to be as follows:

$$\begin{Bmatrix} \sigma_x \\ \sigma_y \\ \tau_{xy} \end{Bmatrix} = \begin{Bmatrix} \overline{Q}_{11} & \overline{Q}_{12} & \overline{Q}_{16} \\ \overline{Q}_{12} & \overline{Q}_{22} & \overline{Q}_{26} \\ \overline{Q}_{16} & \overline{Q}_{26} & \overline{Q}_{66} \end{Bmatrix} \begin{Bmatrix} \varepsilon_x - \overline{\alpha}_1 \Delta T - \overline{\beta}_1 \Delta C \\ \varepsilon_y - \overline{\alpha}_2 \Delta T - \overline{\beta}_2 \Delta C \\ \gamma_{xy} - \overline{\alpha}_6 \Delta T - \overline{\beta}_6 \Delta C \end{Bmatrix}$$

(14)

where the components of the transformed stiffness matrix Q are given by:

$$\overline{Q}_{11} = Q_{11}m^4 + 2(Q_{12} + 2Q_{66})m^2n^2 + Q_{22}n^4$$

$$\overline{Q}_{12} = (Q_{11} + Q_{22} - 4Q_{66})m^2n^2 + Q_{12}(m^4 + n^4)$$

$$\overline{Q}_{22} = Q_{11}n^4 + 2(Q_{12} + 2Q_{66})m^2n^2 + Q_{22}m^4$$

$$\overline{Q}_{66} = (Q_{11} + Q_{22} - 2Q_{12} - 2Q_{66})m^2n^2 + Q_{66}(m^4 + n^4)$$

$$\overline{Q}_{16} = (Q_{11} - Q_{12} - 2Q_{66})m^3n + (Q_{12} - Q_{22} + 2Q_{66})mn^3$$

(15)

$$\overline{Q}_{26} = (Q_{11} - Q_{12} - 2Q_{66})mn^3 + (Q_{12} - Q_{22} + 2Q_{66})m^3n$$

$$\overline{\alpha}_1 = \alpha_1 m^2 + \alpha_2 n^2$$

$$\overline{\alpha}_2 = \alpha_1 n^2 + \alpha_2 m^2$$

$$\overline{\alpha}_6 = (\alpha_2 - \alpha_1)mn$$

$$\overline{\beta}_1 = \beta_1 m^2 + \beta_2 n^2$$

$$\overline{\beta}_2 = \beta_1 n^2 + \beta_2 m^2$$

$$\overline{\beta}_6 = (\beta_2 - \beta_1)mn$$

Inspection of eqs (15) shows that for $\theta = 0°$, $\overline{Q}_{ij} = Q_{ij}$ ($ij = 11, 12, 22,$ and 66) and $\overline{Q}_{16} = \overline{Q}_{26} = 0$. Also, for $\theta = 90°$, $\overline{Q}_{11} = Q_{22}$, $\overline{Q}_{12} = Q_{12}$, $\overline{Q}_{22} = Q_{11}$, $\overline{Q}_{66} = Q_{66}$, $\overline{Q}_{16} = \overline{Q}_{26} = 0$. In both of these cases, $\overline{Q}_{16} = \overline{Q}_{26} = 0$ which means that, as an isotropic material, there is no shear-normal coupling.

If θ is *not* 0, 90°, or $-90°$, or $\overline{Q}_{16}$ and $\overline{Q}_{26}$ are not zero. Thus, shear-normal coupling occurs. This means that a shear strain produces normal stresses and vice versa. Also, a temperature or moisture change produces shear stress. These shear-normal effects constitute the most different behavior of composites, even qualitatively, as compared to isotropic materials.

The effects of fiber orientation on the various thermoelastic quantities for a unidirectional graphite-epoxy composite are shown in Fig. 4.

Special Considerations

For those readers wishing to learn more about three-dimensional anisotropic elastic behavior, reference is made to the books by Hearmon[2] and Lekhnitskii[3,4].

It should be cautioned that, in general, for anisotropic materials (including orthotropic), the principal stress directions (θ_σ) do not coincide with the principal strain directions (θ_ε). For certain materials and orientations, the difference ($\theta_\sigma - \theta_\varepsilon$)

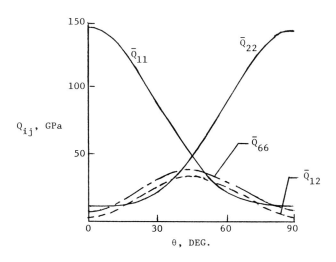

Fig. 4—Effect of fiber orientation on the thermoelastic properties of graphite-epoxy.

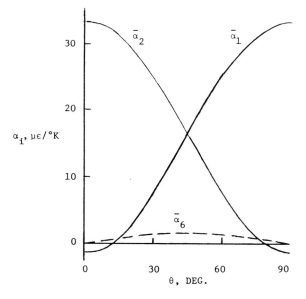

can reach 30 degrees. Fortunately, this does not usually cause a problem, because all of the failure theories proposed for composites relate to strength quantities defined and measured relative to the material-symmetry directions, *not* principal-stress or principal-strain directions; see the textbook by Jones[1], pp. 59-83, for example.

The elastic constants (E's, C's, or Q's) can be predicted analytically or numerically using equations of micromechanics. It is found that they depend not only upon the fiber volume fraction and the properties of the fiber and matrix but also to a lesser extent upon the fiber array and fiber cross-sectional shape.

See, for example, the books by Jones[1] and Christensen[6] and the survey by Bert and Francis[7]. However, for experimental-mechanics use, it is better to measure the macroscopic properties directly from tests on properly designed specimens; see the work of Bert[5,8] and the books by Tarnopol'skii and Kincis[9] and Carlsson and Pipes[10].

The reader is cautioned that the familiar interrelationship among elastic modulus (E), Poisson's ratio (v), and shear modulus (G) for isotropic materials has no counterpart in composites. In other words, in composites, the shear modulus G is an independent elastic constant.

Also, the theoretical limits on Poisson's ratio for isotropic materials (-1 and $+0.5$) do not hold for composites. Jones[1] (pp. 41-45) gives expressions for the appropriate limits for orthotropic materials.

Some composites, especially those with very soft matrices (such as elastomers), exhibit drastically different behavior in tension and compression. See the experimental results published by Patel et al.[11] and Bert and Kumar[12]. For such materials, it is expedient to use a different set of elastic constants depending upon whether the fibers are strained in tension or compression; see material model II in the paper by Bert[13] and the survey papers by Bert[14] and Bert and Reddy[15].

For dynamic loading of materials, it is convenient to approximate the material behavior using the complex modulus approach, i.e., replacing the ordinary stiffnesses by a real storage modulus. For the micromechanics of such a composite, see the work of Hashin[16] and Chang and Bert[17], and the survey by Bert[18]. For appropriate experimental methods, see the work of Bert and Clary[19] and for application to dynamic-response prediction, see Siu and Bert[20].

For nonlinear stress-strain behavior, as in the case of metal-matrix composites; see, e.g., the work of Dvorak[21].

Acknowledgments

The author acknowledges the helpful suggestions of Professors Akhtar S. Khan and Alfred G. Striz of the University of Oklahoma and the skillful typing of Mrs. Rose Benda.

References

1. Jones, R.M., Mechanics of Composite Materials, Scripta Book Co., Washington, DC (1975).
2. Hearmon, R.F.S., An Introduction to Applied Anisotropic Elasticity, Oxford University Press, London (1961).
3. Lekhnitskii, S.G., Theory of Elasticity of an Anisotropic Body, Eng. trans., Holden-Day, San Francisco (1963).

4. Lekhnitskii, S.G., *Theory of Elasticity of an Anisotropic Body*, 2nd Ed., Eng. trans., Mir Publishers, Moscow (1981).

5. Bert, C.W., "Experimental Characterization of Composites," *Composite Materials*, ed. L.J. Broutman and R.H. Krock, Academic Press, New York, Vol. 8: *Structural Design and Analysis*, Part II, ed. C.C. Chamis, Ch. 9, 73-133 (1975).

6. Christensen, R.M., *Mechanics of Composite Materials*, Wiley, New York (1979).

7. Bert, C.W. and Francis, P.H., "Composite Material Mechanics: Thermoelastic Micromechanics," *Transactions, New York Academy of Sciences*, Ser. II, **36**, 663-674 (1974).

8. Bert, C.W., "Static Testing Techniques for Filament-Wound Composite Materials," *Composites* **5**, 20-26 (1974).

9. Tarnopol'skii, Yu.M. and Kincis, T., *Static Test Methods for Composites*, Eng. trans., Van Nostrand Reinhold, New York (1985).

10. Carlsson, L.A. and Pipes, R.B., *Experimental Characterization of Advanced Composite Materials*, Prentice-Hall, Englewood Cliffs, NJ (1987).

11. Patel, H.P., Turner, J.L., and Walter, J.D., "Radial Tire Cord-Rubber Composite," *Rubber Chemistry and Technology*, **49**, 1095-1110 (1976).

12. Kumar, M. and Bert, C.W., "Experimental Characterization of Mechanical Behavior of Cord-Rubber Composites," *Tire Science and Technology*, **10**, 37-54 (1982); addendum, **15**, 68-70 (1987).

13. Bert, C.W., "Models for Fibrous Composites with Different Properties in Tension and Compression," *ASME J. Eng. Mat. and Tech.*, **99H**, 344-349 (1977).

14. Bert, C.W., "Recent Advances in Mathematical Modeling of the Mechanics of Bimodulus, Fiber-Reinforced Composite Materials," *Proc. 15th Ann. Mtg., Soc. Eng. Sci.*, Gainesville, FL, 101-106 (1978).

15. Bert, C.W. and Reddy, J.N., "Mechanics of Bimodular Composite Structures," *Mechanics of Composite Materials: Recent Advances, Proc. IUTAM Symp.*, Blacksburg, VA (1982); Pergamon Press, Oxford, 323-337 (1983).

16. Hashin, Z., "Complex Moduli of Viscoelastic Composites — II. Fiber Reinforced Materials," *Int. J. Solids and Structures*, **6**, 797-807 (1970).

17. Chang, S. and Bert, C.W., "Analysis of Damping for Filamentary Composite Materials," *Composite Materials in Engineering Design, Proc. 6th St. Louis Symp.* (1972); ASM, 51-62 (1973).

18. Bert, C.W., "Composite Materials: A Survey of the Damping Capacity of Fiber-Reinforced Composites," *Damping Application for Vibration Control*, ed. P.J. Torvik, ASME, **AMD 38**, 53-63 (1980).

19. Bert, C.W. and Clary, R.R., "Evaluation of Experimental Methods for Determining Dynamic Stiffness and Damping of Composite Materials," *Composite Materials: Testing and Design (3rd Conference)*, ASTM Spec. Tech. Pub. 546, 250-265 (1974).

20. Siu, C.C. and Bert, C.W., "Sinusoidal Response of Composite-Material Plates with Material Damping," *ASME J. Eng. for Ind.*, **96B**, 603-610 (1974).

21. Dvorak, G.J., "Metal Matrix Composites: Plasticity and Fatigue," *Mechanics of Composite Materials: Recent Advances, Proc. IUTAM Symp.*, Blacksburg, VA (1982); Pergamon Press, Oxford, 73-91 (1983).

Section IIB

Classical Lamination Theory

by Charles W. Bert

Introduction

In a typical structural application of a composite, multiple layers (or laminae) of unidirectional composites are stacked together at various angles to form a laminate. The stacking sequence and orientations of the individual layers give the laminate designer additional 'degrees of freedom' to 'tailor' or optimize the design with respect to strength, stiffness, buckling load, vibration response, panel flutter or other desired performance objective.

The purpose of lamination theory is to predict the behavior of a laminate from a knowledge of the material properties of the individual layers and the laminate geometry. In the next subsection is presented what is generally known as classical lamination theory, usually attributed to Reissner and Stavksy[1]. This is followed by a collection of results for a variety of popular laminate configurations. The final subsection discusses a number of special considerations.

Development of the Theory

Classical lamination theory is based upon the following simplifying engineering assumptions. (1) Each layer is thin and constructed of macroscopically homogeneous, orthotropic, linear-elastic material as discussed in subsection IIA-3. (2) The entire laminate and all of the individual layers are assumed to be in a state of plane stress. (3) The layers are perfectly bonded together. (4) The Kirchhoff hypothesis is invoked, i.e., plane, normal cross sections of the entire laminate before deformation remain plane, normal to the deflected middle surface, and do not change in thickness.

A corollary of hypothesis (4) is that the in-plane displacements vary linearly through the entire thickness of the laminate, while the normal deflection is uniform through the thickness. Thus, the thickness-shear (transverse-shear) strains and the thickness-normal strain are all zero. The only nonzero strains are the in-plane strains (two normal strains ε_1 and ε_2 and one shear strain ε_6) which vary linearly through the thickness:

$$\begin{Bmatrix} \varepsilon_1 \\ \varepsilon_2 \\ \varepsilon_6 \end{Bmatrix} = \begin{Bmatrix} \varepsilon_1^o \\ \varepsilon_2^o \\ \varepsilon_6^o \end{Bmatrix} + z \begin{Bmatrix} \varkappa_1 \\ \varkappa_2 \\ \varkappa_6 \end{Bmatrix} \qquad (1)$$

Here, ε_1^o and ε_2^o are the mid-plane normal strains, ε_6^o is the mid-plane shear strain, $\varkappa_1$ and $\varkappa_2$ are the bending curvatures, $\varkappa_6$ is the twisting curvature, and z is the thickness-direction position coordinate, measured from the mid-plane.

As was shown in Section IIA, the stress state in a typical layer (denoted by the subscript k), with the fibers oriented at an angle θ_k from the reference axes, can be expressed as

$$\begin{Bmatrix} \sigma_1 \\ \sigma_2 \\ \sigma_6 \end{Bmatrix} = \begin{bmatrix} \bar{Q}_{11} & \bar{Q}_{12} & \bar{Q}_{16} \\ \bar{Q}_{12} & \bar{Q}_{22} & \bar{Q}_{26} \\ \bar{Q}_{16} & \bar{Q}_{26} & \bar{Q}_{66} \end{bmatrix}_k \begin{Bmatrix} \varepsilon_1 \\ \varepsilon_2 \\ \varepsilon_6 \end{Bmatrix} \qquad (2)$$

For laminates, as in plate and shell structures, it is usually more convenient to work with the resultant forces and moments, expressed per unit width, than it is to deal with stress components in each individual layer. The in-plane forces per unit width for a general laminate consisting of n plies (see Fig. 1) can be determined by

$$\begin{Bmatrix} N_1 \\ N_2 \\ N_6 \end{Bmatrix} = \int_{-h/2}^{h/2} \begin{Bmatrix} \sigma_1 \\ \sigma_2 \\ \sigma_6 \end{Bmatrix} dz = \sum_{k=1}^{n} \int_{h_k-1}^{h_k} \begin{Bmatrix} \sigma_1 \\ \sigma_2 \\ \sigma_6 \end{Bmatrix} dz \qquad (3)$$

Here, N_1 and N_2 are in-plane normal forces per unit width, and N_6 is the in-plane shear force per unit width (see Fig. 2).

Similarly, the moment resultants or stress couples can be expressed as

$$\begin{Bmatrix} M_1 \\ M_2 \\ M_6 \end{Bmatrix} = \int_{-h/2}^{h/2} z \begin{Bmatrix} \sigma_1 \\ \sigma_2 \\ \sigma_6 \end{Bmatrix} dz = \sum_{k=1}^{n} \int_{h_k-1}^{h_k} z \begin{Bmatrix} \sigma_1 \\ \sigma_2 \\ \sigma_6 \end{Bmatrix} dz \qquad (4)$$

Here, M_1 and M_2 are bending moments per unit width, and M_6 is the twisting moment per unit width (see Fig. 2).

Fig. 1—Geometry of an n-ply laminate

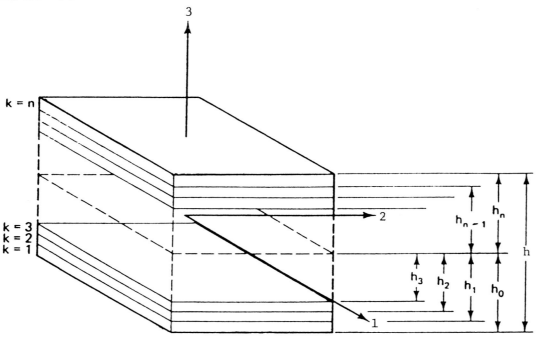

Fig. 2—In-plane forces and moment resultants acting on a laminate

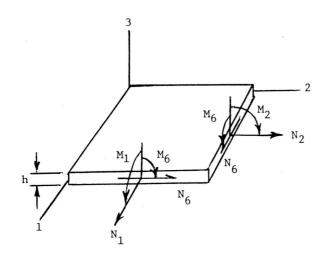

Substituting eqs. (1) and (2) into eq (3), one obtains

$$\begin{Bmatrix} N_1 \\ N_2 \\ N_6 \end{Bmatrix} = \begin{bmatrix} A_{11} & A_{12} & A_{16} \\ A_{12} & A_{22} & A_{26} \\ A_{16} & A_{26} & A_{66} \end{bmatrix} \begin{Bmatrix} \varepsilon_1^0 \\ \varepsilon_2^0 \\ \varepsilon_6^0 \end{Bmatrix} + \begin{bmatrix} B_{11} & B_{12} & B_{16} \\ B_{12} & B_{22} & B_{26} \\ B_{16} & B_{26} & B_{66} \end{bmatrix} \begin{Bmatrix} K_1 \\ K_2 \\ K_6 \end{Bmatrix}$$

(5)

where the stretching stiffnesses are given by

$$A_{ij} = \int_{-h/2}^{h/2} \overline{Q}_{ij}\, dz = \sum_{k=1}^{n} (h_k - h_{k-1})(\dot{Q}_{ij})_k \qquad (6)$$

$$(i,j = 1,2,6)$$

and the bending-stretching coupling stiffness are

$$B_{ij} = \int_{-h/2}^{h/2} z\overline{Q}_{ij}\, dz = \frac{1}{2} \sum_{k=1}^{n} (h_k^2 - h_{k-1}^2)(\overline{Q}_{ij})_k$$

$$(i,j = 1,2,6) \qquad (7)$$

In a similar fashion, substitution of eqs (1) and (2) into eq (4) yields

$$
\begin{Bmatrix} M_1 \\ M_2 \\ M_6 \end{Bmatrix} = \begin{bmatrix} B_{11} & B_{12} & B_{16} \\ B_{12} & B_{22} & B_{26} \\ B_{16} & B_{26} & B_{66} \end{bmatrix} \begin{Bmatrix} \varepsilon_1^0 \\ \varepsilon_2^0 \\ \varepsilon_6^0 \end{Bmatrix} + \begin{bmatrix} D_{11} & D_{12} & D_{16} \\ D_{12} & D_{22} & D_{26} \\ D_{16} & D_{26} & D_{66} \end{bmatrix} \begin{Bmatrix} K_1 \\ K_2 \\ K_6 \end{Bmatrix}
$$

(8)

Here, the B's are as defined in eq (7) and the bending stiffnesses are given by

$$
D_{ij} = \int_{-h/2}^{h/2} z^2 \bar{Q}_{ij}\, dz = \frac{1}{3} \sum_{k=1}^{n} (h_k^3 - h_{k-1}^3)(\bar{Q}_{ij})_k
$$

$$(i,j = 1,2,6)$$ (9)

Equations (5) and (8) can be combined into the following single 6 x 6 matrix equation:

$$
\begin{Bmatrix} N_1 \\ N_2 \\ N_6 \\ M_1 \\ M_2 \\ M_6 \end{Bmatrix} = \left[\begin{array}{ccc|ccc} A_{11} & A_{12} & A_{16} & B_{11} & B_{12} & B_{16} \\ A_{12} & A_{22} & A_{26} & B_{12} & B_{22} & B_{26} \\ A_{16} & A_{26} & A_{66} & B_{16} & B_{26} & B_{66} \\ \hline B_{11} & B_{12} & B_{16} & D_{11} & D_{12} & D_{16} \\ B_{12} & B_{22} & B_{26} & D_{12} & D_{22} & D_{26} \\ B_{16} & B_{26} & B_{66} & D_{16} & D_{26} & D_{66} \end{array} \right] \begin{Bmatrix} \varepsilon_1^0 \\ \varepsilon_2^0 \\ \varepsilon_6^0 \\ K_1 \\ K_2 \\ K_6 \end{Bmatrix}
$$

(10)

Equation (10), which is known as the laminate constitutive equation, can be written in a much more compact form as

$$
\begin{Bmatrix} N \\ \hline M \end{Bmatrix} = \left[\begin{array}{c|c} A & B \\ \hline B & D \end{array} \right] \begin{Bmatrix} \varepsilon^0 \\ \hline K \end{Bmatrix}
$$

(11)

It is emphasized that eq (10) exhibits a variety of different kinds of coupling. First, all of the B's represent bending-stretching coupling in general. Further, all quantities with subscripts 12 involve Poisson coupling due to the presence of nonzero Poisson's ratio. Finally, all quantities with subscripts 16 and 26 involve normal-shear coupling (often called simply 'shear coupling' in the literature). For example, D_{16} represents coupling between bending (M_1 or k_1) and twisting (k_6 or M_6).

The form of the array of stiffnesses (A, B, and D) appearing in eq. (10) simplifies for many common laminate configurations, as described in the next section.

In developing the preceding equations, thermal and hygrothermal effects were neglected, i.e., it was assumed that the isothermal and isomoisture conditions existed. If the temperature and the moisture concentration change, eq (11) must be replaced by

$$
\begin{Bmatrix} N \\ \hline M \end{Bmatrix} = \left[\begin{array}{c|c} A & B \\ \hline B & D \end{array} \right] \begin{Bmatrix} \varepsilon^0 \\ \hline K \end{Bmatrix} + \begin{Bmatrix} N^T \\ \hline M^T \end{Bmatrix} + \begin{Bmatrix} N^M \\ \hline M^M \end{Bmatrix}
$$

(12)

Here, N^T and M^T represent thermal stretching and thermal bending, respectively, while N^M and M^M represent moisture-induced stretching and moisture-induced bending, respectively. They may be calculated in terms of the individual-layer properties as:

$$
\begin{Bmatrix} N_1^T \\ N_2^T \\ N_6^T \end{Bmatrix} = \int_{-h/2}^{h/2} \begin{Bmatrix} \bar{Q}_{11}\bar{\alpha}_1 + \bar{Q}_{12}\bar{\alpha}_2 + \bar{Q}_{16}\bar{\alpha}_6 \\ \bar{Q}_{12}\bar{\alpha}_1 + \bar{Q}_{22}\bar{\alpha}_2 + \bar{Q}_{26}\bar{\alpha}_6 \\ \bar{Q}_{16}\bar{\alpha}_1 + \bar{Q}_{26}\bar{\alpha}_2 + \bar{Q}_{66}\bar{\alpha}_6 \end{Bmatrix} \Delta T\, dz
$$

(13)

$$
\begin{Bmatrix} M_1^T \\ M_2^T \\ M_6^T \end{Bmatrix} = \int_{-h/2}^{h/2} z \begin{Bmatrix} \bar{Q}_{11}\bar{\alpha}_1 + \bar{Q}_{12}\bar{\alpha}_2 + \bar{Q}_{16}\bar{\alpha}_6 \\ \bar{Q}_{12}\bar{\alpha}_1 + \bar{Q}_{22}\bar{\alpha}_2 + \bar{Q}_{26}\bar{\alpha}_6 \\ \bar{Q}_{16}\bar{\alpha}_1 + \bar{Q}_{26}\bar{\alpha}_2 + \bar{Q}_{66}\bar{\alpha}_6 \end{Bmatrix} \Delta T\, dz
$$

and

$$
\begin{Bmatrix} N_1^M, M_1^M \\ N_2^M, M_2^M \\ N_6^M, M_6^M \end{Bmatrix} = \int_{-h/2}^{h/2} (1,z) \begin{Bmatrix} \bar{Q}_{11}\bar{\beta}_1 + \bar{Q}_{12}\bar{\beta}_2 + \bar{Q}_{16}\bar{\beta}_6 \\ \bar{Q}_{12}\bar{\beta}_1 + \bar{Q}_{22}\bar{\beta}_2 + \bar{Q}_{26}\bar{\beta}_6 \\ \bar{Q}_{16}\bar{\beta}_1 + \bar{Q}_{26}\bar{\beta}_2 + \bar{Q}_{66}\bar{\beta}_6 \end{Bmatrix} \Delta C\, dz
$$

(14)

Here, the $\bar{\alpha}$'s and $\bar{\beta}$'s are the transformed thermal-expansion and moisture-expansion coefficients, respectively, ΔT is the temperature change (from a strain-free temperature) and ΔC is the moisture change (from a strain-free moisture).

Stiffnesses of Specific Laminate Configurations

Before listing the simplifications of applying the theory of the preceding subsection to specific laminate configurations, the standard code or abbreviation used to designate stacking sequence will be illustrated by several examples. First, consider

$$[0_2/90_3/30/-30]_s$$

This means that the first 'ply group' or 'sublaminate', starting from the bottom of the laminate, consists of two plies at an orientation of 0 deg, followed by another ply group of three plies at 90 deg, a single layer at 30 deg, and finally another single layer at -30 deg. The subscript S at the closing bracket indicates that the laminate is symmetric about the laminate midplane ($z = 0$). Thus, the upper half of the laminate has a stacking sequence exactly reverse in order of that of the bottom half. Other ways of designating the same laminate described in the preceding paragraph are

$$[0_2/90_3/30/-30/-30/30/90_3/0_2]_T$$

$$[0_2/90_3/30/-30_2/30/90_3/0_2]_T$$

where the subscript T at the end denotes that this is the stacking sequence for the total laminate.

If a laminate consists of layers of two different composites, i.e., graphite-epoxy (for high stiffness) and glass-epoxy (for low cost), it is called an interlaminar hybrid laminate.

Midplane-Symmetric Laminates

A laminate is said to be midplane symmetrically laminated, or simply symmetrically laminated, if, for each and every layer located a certain distance (z) above the midplane of the laminate, there is an identical layer (same properties, orientation and thickness) located at the same distance below the midplane (i.e., $-z$). From the definition of the bending-stretching coupling stiffnesses in eq (7), it is apparent that *all* of these B_{ij} are equal to zero for a symmetric laminate.

Since even a uniform temperature change or a uniform moisture change causes an unsymmetric laminate to deflect or warp out of its plane, it is usually highly desirable from both production and operational viewpoints to have symmetric laminates.

Aligned and Off-Axis Parallel-Ply Laminates

A parallel-ply laminate is simply a laminate in which all of the major material-symmetry axes (fiber directions) of the individual layers have the same orientation. The plies do not have to be identical in thickness. However, since commercially available prepreg composites come in certain standard ply thicknesses, the layers usually are all of the same nominal thickness. A parallel-ply laminate may be an interlaminar hybrid, i.e., it may have so many layers of graphite-epoxy and so many layers of glass-epoxy, provided that the fiber directions of all layers coincide. Of course, if the layers are not arranged symmetrically about the midplane, a hybrid parallel-ply laminate would exhibit bending-stretching coupling, i.e, $B_{ij} \neq 0$.

A parallel-ply laminate is said to be aligned if the material-symmetry axes of the layers coincide with the reference axes of the laminated plate (such as the sides of a rectangular plate), i.e., $\theta = 0$. Such a plate is called specially orthotropic, since the $\overline{Q}_{ij} = Q_{ij}$ and thus, $\overline{Q}_{16} = \overline{Q}_{26} = 0$. Therefore, all elements of the

stiffness submatrices ([A], [B], [D] with subscripts 16 and 26) are equal to zero. If $\theta \neq 0$, a parallel-ply laminate is said to be an off-axis one. In this case, the stiffness-matrix elements with subscripts 16 and 26 are not zero and the plate is called generally orthotropic. The only purpose of using parallel-ply laminates is to build up a thick plate, since individual plies are usually quite thin (less than 0.010 in.).

Cross-Ply Laminates

A cross-ply laminate is one consisting of an arbitrary number of plies, some of which are oriented at 0 deg and the rest at 90 deg. A regular cross-ply laminate is one having an arbitrary number of plies of the same material and thickness, but with alternating orientations of 0 and 90 deg to the geometric reference axes (plate edges). Thus, by definition, each ply is specially orthotropic; thus, $\overline{Q}_{16} = \overline{Q}_{26} = 0$ for all layers and all elements of the stiffness matrices having subscripts 16 and 26 are equal to zero.

A regular cross-ply laminate having an odd number of layers is automatically symmetric about the midplane. (This is why plywood manufacturers normally produce plywood with only an odd number of plies.) However, if the number of plies is even, two elements of the [B] submatrix, B_{11} and B_{22}, are *not* zero. Specific expressions, obtainable from the general equations, (6, 7 and 9), were presented by Tsai[2] and Bert[3]. In a regular cross-ply laminate with an odd number of layers, the stretching and bending stiffnesses associated with the directions 1 and 2 generally are not equal (unless the layers are all of isotropic materials, of course), i.e., $A_{11} \neq A_{22}$ and $D_{11} \neq D_{22}$. To make $A_{11} = A_{22}$, one could use a ply group of two layers at 90 deg at the center and individual layers at 0 deg at the top and bottom. However, even then $D_{11} \neq D_{22}$, due to the different z-coordinate positions of the 0- and 90-deg layers. Also, it is cautioned that just because $A_{11} = A_{22}$, such a laminate is *not* inplane isotropic, since $A_{66} < (1/2)(A_{11} - A_{12})$.

Angle-Ply Laminates

An angle-ply laminate consists of layers oriented at one or more sets of angles, say $+\theta_1$, $-\theta_1$ and θ_2, $-\theta_2$. In the tire industry, this is called a bias-ply layup.

A regular angle-ply laminate consists of an arbitrary number of layers (n), identical in thickness and material but having alternating orientations of $-\theta$ and $+\theta$ (see Fig. 3). If a regular angle-ply laminate has an odd number of layers, it is a symmetric laminate, so that all of the *B*'s are equal to zero. Unfortunately, however, A_{16}, A_{26}, D_{16}, and D_{26} are not zero; thus, there is in-plane shear-normal coupling and bending-twisting coupling. A regular angle-ply laminate with an even number of layers has $A_{16} = A_{26} = A_{16} = D_{26} = 0$; but unfortunately it is

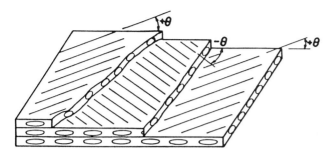

Fig. 3—A typical angle-ply laminate

(1) The total number of layers n must be at least three. (2) The individual layers must have identical orthotropic elastic coefficients (Q_{ij}) and thickness. (3) Each layer in a ply group (or set), denoted by index K, must be oriented at an angle $\theta_K = \pi(K-1)/S$ radians with respect to a reference direction, where S is the number of such ply groups (see Fig. 4).

Since a laminate made according to the Werren-Norris configuration is isotropic in regard to stretching only (submatrix [A]) and not, in general, in regard to [B] and [D], such a laminate is called quasi-isotropic. The simplest example of a quasi-isotropic laminate is a three-layer one with $\theta = 0$, $\pi/3$, $2\pi/3$ radians, i.e., $[0/60/-60]_T$. However, such a laminate is

not symmetric. In fact, since the only B's which are not zero in such a laminate are B_{16} and B_{26}, it is often called an antisymmetric angle-ply laminate.

In regular angle-ply laminates with either an odd or even number of layers, if there are 'many' layers ($n > 10$ or 20), A_{16}, A_{26}, D_{16}, D_{26}, B_{16}. and B_{26} are all either identically zero or negligible. However, it is not always practicable to have many layers, for example, in passenger-car tires. One way to have $A_{16} = A_{26} = B_{16} = B_{26} = 0$, as well as small D_{16} and D_{26} is to use an even number of plies with a two-layer ply group at the center. For example, the simplest configuration of this class, called SBAP (symmetric, balance* angle-ply) by Bert[4], is a four-ply laminate designated as $[\theta/-\theta/-\theta/\theta]_T$, $[\theta/-\theta_2/\theta]_T$, or $[\theta/-\theta]_S$. This configuration is very beneficial for tires, since it eliminates the B_{16} and B_{26} stiffnesses which produce an undesirable characteristic known as 'ply steer', (see Ref. 5).

Quasi-Isotropic Laminates

In certain applications, such as thermally loaded structures and balanced-biaxial-stressed structures, it is desirable that the in-plane behavior of the laminate be isotropic, i.e., the stretching stiffness matrix contain only two independent parameters. If these two are taken to be A_{11} and A_{12} then in-plane isotropy requires that these three conditions be met:

$$A_{22} = A_{11}$$

$$A_{66} = (1/2)(A_{11} - A_{12})$$

$$A_{16} = A_{26} = 0 \qquad (15)$$

The conditions of eqs (15) can be met by a laminate configuration first suggested by Werren and Norris[6]:

Fig. 4—Two examples of quasi-isotropic laminate configurations

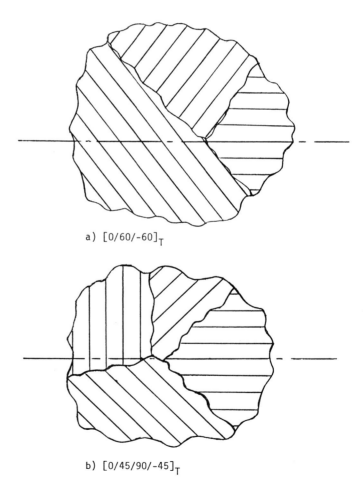

a) $[0/60/-60]_T$

b) $[0/45/90/-45]_T$

---

* *As used in the U.S., the term balanced means the number of plies at $+\theta$ is balanced by those at $-\theta$, so that $A_{16} = A_{26} = 0$.*

not a symmetric laminate. To make it symmetric would require six layers [0/60/-60]$_s$, i.e., [0/60/-60/-60/60/0]$_T$. A more popular quasi-isotropic laminate configuration is [0/45/-45/90]$_s$ or [0/90/45/-45]$_s$ which contains a total of eight layers. Ref. 3 provides explicit equations for all of the laminate stiffnesses.

[0$_m$/ ± θ$_n$]$_s$ Laminates

A very popular lamination scheme for structural panels in the airframe industry is [0$_m$/ ± θ$_n$]$_s$, where, of course, m and n are integers. There are several advantages of this configuration over other simpler ones: (1) Unlike a regular angle-ply laminate, there is a reserve of strength after first-ply failure. This reserve is analogous to the reserve of strength provided by plasticity in isotropic materials. (2) It is possible to tailor the various properties continuously rather than in discrete steps as is required in cross-ply laminates.

Special Considerations

In the foregoing subsections, it was assumed that the material was linear elastic, thin, and not subject to delamination. If any of these hypotheses are violated, it is obvious that the composite structure will not behave as predicted. However, as was first shown by Tsai[2], for instance, the agreement between theory and experiment for the laminate stiffnesses, [A], [B], and [D], was quite good for thin laminates undergoing small deflections.

If the laminate is not thin, it is necessary to consider the transverse shear flexibility (see both Whitney and Pagano[7] and Whitney's[8] recent book. If delamination is a concern, not only transverse shear flexibility but thickness-normal stiffness may be important. If the laminate is relatively thin and the out-of-plane loading is sufficiently high, the laminate may undergo large enough slopes and deflections to induce membrane action and thus geometric nonlinearity. (See the book by Chia[9] for an extensive discussion of this topic.)

Considerable care must be exercised in the design of appropriate test specimens of laminated material if the test objective is to obtain the ply constitutive relations. For example, a thin-walled torsion tube is, in general, an excellent specimen in which to determine the ply shear modulus (G_{12}) of a filament-wound composite. However, from the transformation equations for shear modulus (see Section IIA), it can be shown that if the ply angles are ±45 deg, the measured torque vs angle relation is completely independent of G_{12}!

Some materials have drastically different elastic properties depending upon whether or not the fibers are stretched or compressed; this is the case for tire cord-rubber, for example. Then, the material must be modeled as bimodular, and the analysis presented in Section IIB-2 would not be appropriate. (See the work of Bert and Reddy[10] for an extensive derivation for this case).

Acknowledgments

The author acknowledges helpful suggestions of his colleague, Professor Akhtar S. Khan, and the skillful typing of Mrs. Rose Benda.

References

1. Reissner, E. and Stavsky, Y., "Bending and Stretching of Certain Types of Heterogeneous Aeolotropic Elastic Plates," ASME J. Appl. Mech., 28, 402-408 (1961),
2. Tsai, S.W., "Structural Behavior of Composite Materials," NASA CR-71 (July 1964).
3. Bert, C.W., "Analysis of Plates," Composite Materials, ed. L.J. Broutman and R.H. Krock, Academic Press Network, Vol. 7: Structural Design and Analysis, Part I, ed. C.C. Chamis, Ch.4, 149-206 (1975).
4. Bert, C.W., "Optimal Design of a Composite-Material Plate to Maximize Its Fundamental Frequency," Journal of Sound and Vibration, 50, 229-237 (1977).
5. Bert, C.W., "Simplified Prediction of Ply Steer in Radial Tires," Tire Sci. and Tech., 8, 3-9 (1980).
6. Werren, F. and Norris, C.B., "Mechanical Properties of a Laminate Designed to be Isotropic," Forest Products Laboratory, Madison, WI, Rep. No. 1841 (1953).
7. Whitney, J.M. and Pagano, N.J., "Shear Deformation in Heterogeneous Anisotropic Plates," ASME J. Appl. Mech., 37, 1031-1036 (1970).
8. Whitney, J.M., Structural Analysis of Laminated Anisotropic Plates, Technomic, Lancaster, PA, Ch. 10 (1987).
9. Chia, C.Y., Nonlinear Analysis of Plates, McGraw-Hill, New York, Ch. 5 and 7 (1980).
10. Bert, C.W. and Reddy, J.N., "Bending of Thick Rectangular Plates Laminated of Bimodulus Composite Materials," AIAA Journal 19, 1342-1349 (1981).

Section IIIA

Fundamental Strain-Gage Technology

by M.E. Tuttle

Introduction

In 1856 Lord Kelvin reported that the resistance of copper and iron wires increased when subjected to a tensile strain[1]. This basic discovery has ultimately led to the development of the modern resistance foil strain gage. The 'birthdate' of the strain gage is considered to be 1938, when Ruge and Simmons, working independently, each measured strains using bonded-wire strain gages. The underlying concept of resistance strain gages is very simple. In essence, an electrically conductive wire is securely bonded to a structure of interest, and the resistance of the wire is measured before and after the structure is loaded. Since the wire is firmly bonded to the structure, strains induced in the structure are also induced in the wire. This results in a change in wire resistance, which serves as an indirect measure of the strain induced in the structure.

Although the underlying concept is simple, there are several sources of error which may lead to erroneous strain measurements if not properly accounted for. These sources of error can be loosely grouped into the following six categories:

(a) The wire must be firmly bonded to the structure, so that the deformation of the wire is an accurate reflection of the deformation of the structure.

(b) The wire must not locally reinforce the structure, otherwise the strain field in the vicinity of the wire will be disturbed and an inaccurate measure of strain will be obtained.

(c) The wire must be electrically insulated from the structure.

(d) The change in wire resistance per unit microstrain is very small; for a 350 Ω gage the change in resistance is on the order of 0.0007 $\Omega/\mu\varepsilon$. This small change in resistance must nevertheless be measured accurately if an accurate measure of strain is to be obtained.

(e) The structure (as well as the wire) may be deformed by mechanisms other than an applied load. The most common example is a change in temperature. If a structure composed of a homogeneous material is free to expand or contract, then a uniform change in temperature will cause the structure to uniformly expand or contract. This gives rise to 'apparent thermal strains', which are not associated with stresses induced within the structure, and are generally not of interest.

(f) The resistance of the wire may be changed by mechanisms other than physical deformation. For example, an aggressive environment may cause oxidation of the wire, causing a change in wire resistance and hence an erroneous strain reading.

Most of the strain-gage technologies discussed in this manual (or elsewhere[1,2]) are directed toward avoiding one or more of the potential sources of error listed above.

The original bonded strain gages developed by both Ruge and Simmons consisted of several 'loops' of a continuous wire; several loops were used to increase the sensitivity of the gage. However, the wire strain gage has been largely displaced by the metal-foil strain gage, which was introduced in the mid-1950s. Foil strain gages are produced from a parent metal foil using a photoetch process. The foil gage is normally bonded to a thin polymeric backing material, and the entire assembly (i.e., the strain gage) is adhesively bonded to the test structure. Foil gages are generally preferred over wire gages due to improvements in gage resistance and sensitivity tolerances, a decrease in gage thickness, and because of the wide variety of grid shapes that can easily be produced using the photoetch process. Although the vast majority of strain gages in use today are metal-foil gages, wire gages are still used for measurements at high temperatures ($\approx$ 400°-1100°C) and other special-purpose applications.

This article is intended as a general introduction to resistance foil strain gages. A few examples of the unusual effects of orthotropic material behavior on strain gage measurements are also discussed. In following articles in this manual the application of strain gages to polymeric composite materials is discussed directly and in detail.

Strain-Gage Calibration

Calibration of a strain gage is performed by the gage manufacturer, and is reported to the user in the form of four gage characteristics: gage resistance, gage factor, transverse-sensitivity coefficient, and self-temperature-compensation (S-T-C) number. These parameters are measured for a statistical number of gages from each lot produced, and the measured values are included with each strain-gage package purchased by the consumer. Note that these parameters are *not* measured for the individual gages supplied to the user; the values reported are average values for the entire strain-gage lot. When placing an order, the consumer specifies the desired gage resistance and S-T-C number. The gage factor and transverse sensitivity vary from lot to lot, and (in general) are not specified by the consumer.

Note that the 'calibration' process being described herein is calibration of the strain gage itself. Calibration of the *total* strain-gage installation (including the effects of leadwire resistance, gage-circuit linearity, amplifier linearity, etc.) must be performed by the strain-gage user. These latter aspects of strain-gage calibration have been described elsewhere.[1,2]

Strain-Gage Resistance

During the strain measurement process, an excitation voltage is applied to the strain gage, causing an electrical current to flow through the gage. Most of the power applied to the gage is dissipated in the form of heat, which must be conducted away from the gage site by the underlying specimen substrate. In general, the greater the voltage applied, the greater the sensitivity of strain measurement. However, the power applied to a strain gage must be below some maximum level, otherwise excessive heat is generated which cannot be adequately dissipated by the substrate, and the gage performance is adversely affected[3]. The appropriate power level is a function of many variables, including gage alloy, gage backing and encapsulation materials, adhesive used, and substrate material. For a given power level, a relatively high gage resistance is desired so as to reduce the level of current and amount of heat generated, resulting in acceptable strain-gage stability.

Strain gages with a resistance of 120, 350, or 1000 Ω are most commonly used and are most readily available. Selection of the appropriate power level and strain-gage resistance for general-purpose strain-gage work is discussed in Ref. 3. Selection of strain gages specifically for use with composites is discussed in a following chapter, "Strain Gages on Composites—Gage-Selection Criteria." For a 350 Ω strain gage used in a Wheatstone bridge circuit[1,2], excitation voltages ranging from about 2 to 4 V are typical.

Gage Factor and Transverse-Sensitivity Coefficient

In essence, the 'gage factor' is a measure of the sensitivity of the strain gage to strains acting in the direction of the gage grid, while the 'transverse sensitivity' is a measure of the sensitivity of the gage to strains acting transverse to the gage grid. In most applications a strain gage is used to measure strains in the grid direction, and hence a very low transverse sensitivity is usually desirable. Although the terms 'gage factor' and 'transverse sensitivity' refer to the gage response to two different strains, the standard calibration procedures which have evolved to measure these parameters have intimately linked the two terms. The relation between 'gage factor' and 'transverse-sensitivity coefficient' will be discussed below, and must be thoroughly understood in order to properly interpret strain-gage measurements.

Fig. 1—A resistance foil strain gage subjected to a biaxial strain field (schematic)

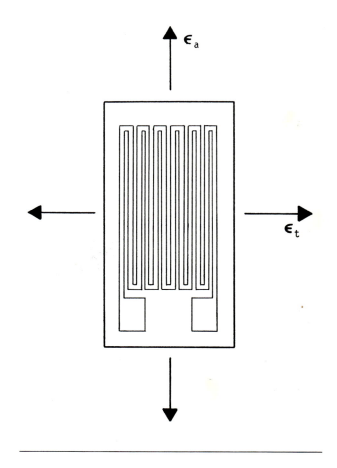

A strain gage subjected to a biaxial-strain field is shown in Fig. 1. Assuming temperature remains constant, the change in gage resistance induced by the biaxial-strain field is given by:

$$\frac{\delta R}{R} = F_a \varepsilon_a + F_t \varepsilon_t \qquad (1)$$

where:

R	=	original gage resistance
δR	=	change in gage resistance
ε_a	=	strain in gage grid direction
ε_t	=	strain transverse to gage grid
F_a	=	axial-gage factor
F_t	=	transverse-gage factor

(Note that neither the 'axial-gage factor' nor the 'transverse-gage factor' which appear in eq (1) are equal to the 'gage factor,' F_g, reported by the manufacturer.)

The transverse sensitivity coefficient, K, is defined as:

$$K \equiv \frac{F_t}{F_a} \qquad (2)$$

This coefficient is one of the calibration parameters supplied by the manufacturer with each strain-gage package, and is measured using a standard testing procedure[4]. The value of K is normally within a range of about -0.05 to 0.05, which indicates that a strain gage is typically much less sensitive to strains acting transverse to the grid than to those acting in the grid direction. Note that gage manufacturers customarily report K as a percentage value, so that the range of K as reported by the manufacturers is from about -5 to 5 percent. Equation (1) can be rewritten in terms of K as follows:

$$\frac{\delta R}{R} = F_a(\varepsilon_a + K\varepsilon_t) \qquad (3)$$

During calibration the gage is mounted to a standard calibration material and subjected to a uniaxial-stress field[4]. The grid direction is orientated parallel to the uniaxial stress. *Under these conditions,* the transverse strain applied to the gage is due to the Poisson effect, and is given by

$$\varepsilon_t = -\mu_o \varepsilon_a \qquad (4)$$

where

μ_o	=	Poisson's ratio of the standard calibration material used by the manufacturer (normally $\mu_o = 0.285$)

Equation (3) can be rewritten *for this loading condition* as:

$$\frac{\delta R}{R} = F_a (1 - \mu_o K)\varepsilon_a \qquad (5)$$

The 'gage factor' supplied by the manufacturer is defined as

$$F_g \equiv F_a(1 - \mu_o K) \qquad (6)$$

or, equivalently,

$$F_g \equiv F_a - \mu_o F_t$$

Modern foil strain gages exhibit a gage factor of about 2.00 ± 0.20. Finally, eq (5) can be rearranged as:

$$\varepsilon_a = \frac{\left[\dfrac{\delta R}{R}\right]}{F_g} \qquad (7)$$

Equation (7) has been derived based upon the following assumptions: (1) The strain gage is subjected to a uniaxial-stress field; (2) The gage grid is parallel to the direction of stress; (3) The gage is mounted on a material whose Poisson's ratio equals μ_o. If a strain gage is used during a test in which all three of these conditions hold, then the measured strain ε_m is given by eq (7) directly, i.e.,

$$\varepsilon_m = \varepsilon_a = \frac{\left[\dfrac{\delta R}{R}\right]}{F_g} \qquad (8)$$

Under any other conditions eq (8) is not valid, and in some cases appreciable measurement error can occur due to transverse sensitivity. In these cases the measured strain must be corrected for transverse-sensitivity effects. At least two orthogonal strain measurements are required to correct these errors. A 'biaxial-strain-gage rosette' (i.e., two strain gages mounted so as to measure two orthogonal normal strains) is shown schematically in Fig. 2(a). Denoting the measured strains as ε_{mx} and ε_{my}, the true strains in the x and y directions, ε_x and ε_y, are given by:

$$\varepsilon_x = \frac{(1 - \mu_o K)(\varepsilon_{mx} - K\varepsilon_{my})}{1 - K^2}$$

$$\varepsilon_y = \frac{(1 - \mu_o K)(\varepsilon_{my} - K\varepsilon_{mx})}{1 - K^2} \qquad (9)$$

Equations (9) are the transverse-sensitivity correction equations for use with biaxial-strain-gage rosettes. Two other gage rosettes are commonly available: a

Fig. 2—Commonly available strain gage rosettes

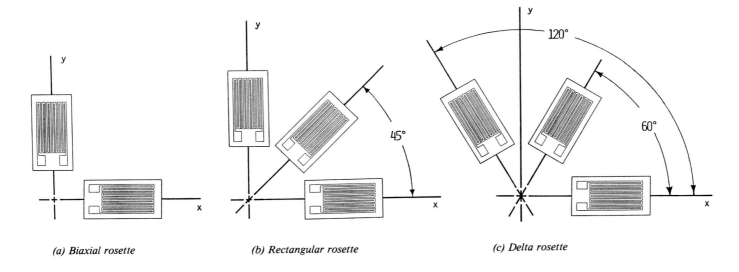

| (a) Biaxial rosette | (b) Rectangular rosette | (c) Delta rosette |

'rectangular' three-element rosette [Fig. 2(b)], and a 'delta' three-element rosette [Fig. 2(c)]. The correction equations for use with these gage configurations are as follows. For rectangular rosettes, denoting the measured strains as ε_{mx}, $\varepsilon_{m45°}$, and ε_{my}, the true strains ε_x, $\varepsilon_{45°}$, and ε_y are given by:

$$\varepsilon_x = \frac{(1 - \mu_o K)\,(\varepsilon_{mx} - K\varepsilon_{my})}{1 - K^2}$$

$$\varepsilon_{45°} = \frac{(1 - \mu_o K)}{1 - K^2}\left[\varepsilon_{m45°} - K(\varepsilon_{mx} + \varepsilon_{my} - \varepsilon_{m45°})\right] \quad (10)$$

$$\varepsilon_y = \frac{(1 - \mu_o K)\,(\varepsilon_{my} - K\varepsilon_{mx})}{1 - K^2}$$

For delta rosettes, denoting the measured strains as ε_{mx}, $\varepsilon_{m60°}$, and $\varepsilon_{m120°}$, the true strains ε_x, $\varepsilon_{60°}$, and $\varepsilon_{120°}$ are given by:

$$\varepsilon_x = \frac{(1 - \mu_o K)}{1 - K^2}\left[\left[1 + \frac{K}{3}\right]\varepsilon_{mx} - \frac{2}{3}K\left[\varepsilon_{m60°} + \varepsilon_{m120°}\right]\right]$$

$$\varepsilon_{60°} = \frac{(1 - \mu_o K)}{1 - K^2}\left[\left[1 + \frac{K}{3}\right]\varepsilon_{m60°} - \frac{2}{3}K\left[\varepsilon_{mx} + \varepsilon_{m120°}\right]\right]$$

(11)

$$\varepsilon_{120°} = \frac{(1 - \mu_o K)}{1 - K^2}\left[\left[1 + \frac{K}{3}\right]\varepsilon_{m120°} - \frac{2}{3}K\left[\varepsilon_{mx} + \varepsilon_{m60°}\right]\right]$$

In eqs. (9)-(11) it is assumed that the transverse-sensivity coefficient is identical for all gages within the rosette; see Ref. 5 for correction equations allowing an independent value of K for each gage element.

Self-Temperature Compensation Number (STC Number)

It has been previously noted that a structure may be deformed by mechanisms other than mechanical loads. The most common example is a change in temperature. A change in temperature will tend to cause a structure to expand or contract, which may result in 'apparent' thermal-strain measurements. Apparent thermal strains can be very large (200 $\mu m/m/°C$ or greater), and if not properly accounted for can completely obliterate the strain signal associated with stresses within the structure.

Apparent thermal strains arise due to (a) differences in the thermal-expansion coefficients of the test material, gage backing, and gage-foil alloy, and (b) changes in the electrical properties of the gage alloy which occur with a change in temperature. An ideal way of eliminating apparent thermal strains would be to develop a gage alloy with the same ex-

20

pansion coefficient as the test material, and whose electrical properties are independent of temperature. Strictly speaking, this has not been found to be possible. However, by manipulating *both* the expansion coefficient and the electrical properties of the gage alloy (using proprietary alloying and thermal treatment processes), gages have been developed whose expansion coefficients are *in effect* approximately equal to that of the test material. This approach is effective over a specified temperature range, typically from about -50 to 200°C (-60 to 400°F). The match between the effective gage-expansion coefficient and test-material-expansion coefficient is not exact, nor is the difference between the two a linear function of temperature. Typically, apparent thermal-strain levels are reduced to about $\pm 100 \mu m/m$, over the temperature range listed above.

A strain gage which is processed in this manner is said to be 'self-temperature-compensated', and is assigned an 'S-T-C' number which represents the *effective* expansion coefficient of the gage. In practice, one selects the gage S-T-C number which most closely matches the thermal expansion coefficient of the test material. For example, the expansion coefficients of steel and aluminum alloys are approximately 6 $\mu in./in./°F$ and 13 $\mu in./in./°F$, respectively. Therefore gages with S-T-C numbers of 06 and 13 are selected for use with steel and aluminum, respectively.

Although the S-T-C thermal compensation technique is widely employed when using strain gages mounted on isotropic materials, this method is not very well suited for use with composites. Some of the difficulties encountered are:

(a) The thermal expansion of composites is in general a highly orthotropic material property. For example, the in-plane expansion coefficients for a unidirectional graphite-epoxy laminate are about -1.8 $\mu m/m/°C$ (-1.0 $\mu in./in./°F$) in the fiber direction, and about +27 $\mu m/m/°C$ (+15 $\mu in./in./°F$) in the direction transverse to the fibers. This implies that the appropriate gage S-T-C number depends on both material type and gage orientation with respect to the fiber direction.

(b) The thermal-expansion coefficients of polymeric composites often vary from lot to lot, and even more so from manufacturer to manufacturer.

(c) The thermal-expansion coefficients of polymeric composites depend on previous thermal and moisture history.

(d) The effective thermal-expansion coefficient(s) of a multi-angle composite laminate depends upon the specific layup used, and may be varied over a wide range of values.

The above factors severely restrict the S-T-C method of temperature compensation when applied to composites. It is interesting to note that even if these difficulties were overcome, thermal compensation may still not be satisfactory due to transverse-sensitivity effects. That is, the gage has been calibrated for an assumed isotropic expansion or contraction, whereas the composite expands or contracts orthotropically. This results in a transverse strain being applied to the gage, which may have to be accounted for due to transverse-sensitivity effects.

Since the S-T-C method of temperature compensation is not widely used with polymeric composites, selection of the S-T-C number often becomes somewhat arbitrary. Nevertheless, one must specify an S-T-C number when ordering a strain gage. Assuming the S-T-C method will not be used to achieve temperature compensation, it is good practice to simply specify a gage with a readily available S-T-C number, which will in most cases reduce gage delivery times.

It should be emphasized that although the S-T-C method of temperature compensation is not widely used with composites, it is nevertheless *essential* that temperature compensation be achieved. Fortunately, there are alternate ways of eliminating apparent thermal strains[1,2]. Two of these are the 'dummy gage' technique, or use of a 'precalibration' curve. Both of these will be discussed in a following chapter, "Strain Gages on Composites—Temperature Compensation."

Finally, it should be noted that in the above discussion it has been assumed that uniform temperature changes occur during a test, and that the test structure is free to expand or contract. If a nonuniform temperature distribution exists within the structure, or if the structure is constrained so as to restrict natural expansion or contraction, then thermal stresses will be induced. These thermal stresses act in addition to stresses resulting from external loading, and will contribute towards failure of the structure. The objective during temperature compensation is to remove the unwanted strain signal due to apparent thermal-strain effects, but not to remove the strain signal due to thermally or mechanically induced stresses.

Special Considerations When Measuring Strain in Composites

Most practical engineering experience is based upon the familiar behavior of homogeneous isotropic materials. A vast body of literature exists which describes the application of strain gages (and other strain-measurement devices) to such materials. An engineer or technician who has used strain gages to study the behavior of isotropic materials will naturally have developed various 'rules of thumb' based upon this experience.

In contrast, advanced composites are highly hetrogeneous and orthotropic, and have only recently been introduced as load-bearing structural materials. Consequently composites can exhibit surprising and unusual behavior, which would not be expected based upon experience with isotropic materials. In some cases these 'surprises' can lead to erroneous interpretation of experimental results. The 'moral of this story' is that all experimental rules of thumb must be reexamined when dealing with composite materials. Three examples of 'surprising' behavior are discussed below to emphasize this point. The specific examples are (a) errors due to slight gage misalignment, (b) the enhancement of transverse sensitivity errors due to orthotropic material properties, and (c) strain measurements near a free edge.

Fig. 3—A misaligned biaxial rosette mounted on a uniaxial steel specimen (β = misalignment angle)

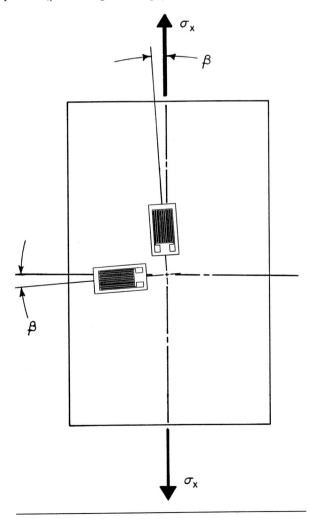

Fig. 4—A misaligned biaxial rosette mounted on an off-axis composite specimen (β = misalignment angle; Θ = Fiber Angle)

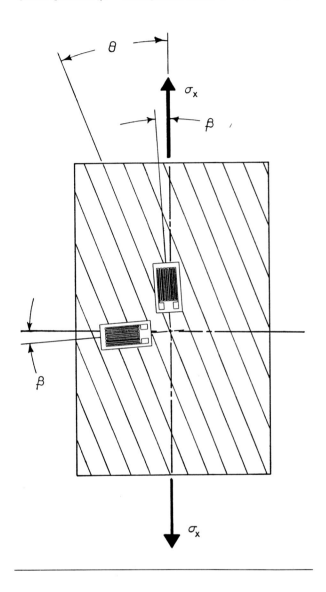

Measurement Error Due To Gage Misalignment

A strain-measurement error occurs whenever a strain gage (or, for that matter, any strain-measurement device) is inadvertently misaligned with respect to the intended direction of strain measurement. This is true regardless of the material being tested. For a single strain gage mounted in a biaxial-strain field, the magnitude of measurement error depends upon three factors[6]: (1) the magnitude of the misalignment error, β, where β equals the angle between the gage axis after bonding and the intended axis of strain measurement, (2) the ratio of the algebraic maximum and minimum principal strains, and (3) the angle ϕ between the maximum-principal-strain axis and the intended axis of strain measurement.

As a sample case, consider a biaxial rosette mounted on a steel tensile specimen as shown schematically in Fig. 3. The rosette is assumed to have been inadvertently misaligned by an angle β. Assumed material properties for steel are listed in Table 1. Referring to the three factors listed above, note that in this case (a) the misalignment error is β, (b) the ratio of the algebraic maximum to minimum principal strain equals $-(1/\mu) = -(1 / 0.285) = 3.51$, and (c) the angle $\phi = 0$, since the principal-stress and strain axes coincide and the intended axes of strain measurement are the principal-strain axes. Using the material properties given in Table 1, it can be shown that percentage errors in axial-strain measurement due to a gage misalignment of $\beta = \pm 2, \pm 4$ deg equal -0.16%, -0.63%, respectively. The percentage errors in transverse strain measurement due to misalignment $\beta = \pm 2, \pm 4$ deg equal -0.55%, -2.2%, respectively. These low percentage errors indicate that a gage-alignment tolerance of ± 4 deg is satisfactory for a uniaxial steel test specimen (although closer gage-alignment tolerances would obviously be good practice!).

TABLE 1—ASSUMED MATERIAL PROPERTIES FOR STEEL AND GRAPHITE/EPOXY

Material	E_1 (GPa)	E_2 (GPa)	μ_{12}	G_{12}(GPa)
Steel	207	---	0.285	80.5
Graphite/ Epoxy	170	8	0.32	6

Now consider the off-axis graphite/epoxy tensile specimen with biaxial rosette shown in Fig. 4. Assumed material properties for graphite/epoxy are listed in Table 1. Referring to the three factors listed above, note that in this case (a) the misalignment error is β, (b) the ratio of the algebraically maximum to minimum strain depends upon the fiber angle Θ[7,8], and ranges from -3.33 to -33.3, and finally (c) the angle ϕ does not (in general) equal zero, since the principal stress and strain axes do not (in general) coincide[9]. Thus, factors (b) and (c) are very much different in the case of a uniaxial composite specimen than in the case of a uniaxial steel specimen.

Typical errors resulting from gage misalignments on unidirectional graphite/epoxy laminates are summarized in Figs. 5 and 6. These results were calculated using the approach described in Ref. 10. In Fig. 5(a) the error in axial strain measurement for an assumed stress level of 60 MPa (8700 psi) is

presented as a function of fiber angle Θ, for misalignment errors of ± 4 and ± 2 deg. The corresponding percentage errors in axial strain measurement are shown in Figure 5(b). As indicated, appreciable percentage errors occur for fiber angles ranging from about 3 to 40 deg; for $B = \pm 4$ deg a maximum error of $\pm 17\%$ occurs at $\Theta \approx 10$ deg. The analogous results for the transverse gage are presented in Figs. 6(a) and (b). In this case appreciable percentage errors in transverse-strain measurements occur for

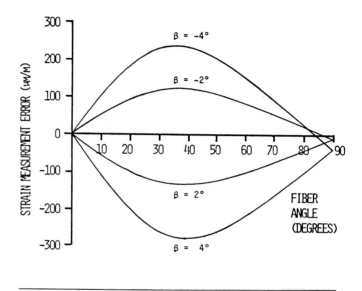

Fig. 5 (a)—Error in axial strain measurement due to misaligned strain gage as a function of fiber angle ($\sigma_x = 60$ MPa)

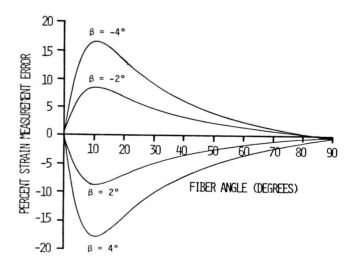

Fig. 5 (b)—Percentage error in axial strain measurement due to misaligned strain gage as a function of fiber angle

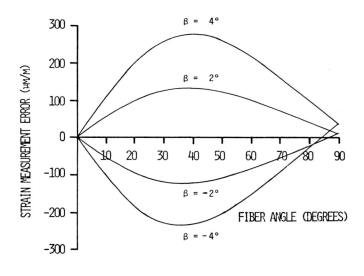

Fig. 6 (a)—Error in transverse strain measurement due to misaligned strain gage as a function of fiber angle (σ_x = 60 MPa)

Fig. 6 (b)—Percentage error in transverse strain measurement due to misaligned strain gage as a function of fiber angle

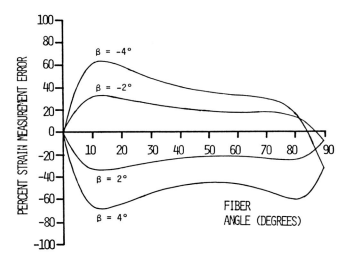

fiber angles ranging from about 2 to 90 deg, with a maximum error of about ±65 percent occurring at $\Theta \approx 14$ degs. Note that the same ±4-deg. gage misalignment which results in a measurement error of 2 percent or less in a steel specimen can produce a measurement error of 65 percent in an off-axis composite specimen! This example demonstrates that for composite materials, very close gage-alignment tolerances are often required. The need for these close tolerances would not be anticipated based upon previous experience with isotropic materials.

Transverse Sensitivity Errors

As described in a preceding section, errors due to gage transverse sensitivity are present in any strain-gage measurement unless (a) the gage is subjected to a uniaxial-stress field, (b) the major axis of the gage is orientated parallel to the applied stress, and (c) the gage is mounted on a material whose Poisson's ratio equals μ_o (normally, $\mu_o = 0.285$). Now, in the general application of strain gages it is often the case that all three of these conditions are violated, and yet errors due to transverse sensitivity are still very low (it should be emphasized that this is often the case, not always the case!). The reasons for this happy circumstance are that gage manufacturers have been successful in reducing the value of the transverse sensitivity coefficient to very low levels (typically less than ±0.05), and that Poisson's ratio for most common structural materials is relatively close to μ_o.

Since transverse-sensitivity errors are so often negligibly small, many engineers simply ignore this source of potential measurement error. However, the orthotropic nature of composites results in a propensity towards transverse-sensitivity error which would not be expected based upon experience with isotropic materials. This enhancement of transverse-sensivity effects can be traced to the fact that the effective Poisson's ratio of a composite is often very much different than μ_o.

To illustrate this point, again consider a biaxial rosette mounted on a steel uniaxial-tensile specimen as shown in Fig. 3 (the rosette is now considered to be perfectly aligned, so that $\beta = 0$). Assuming a transverse-sensitivity coefficient $K = 0.03$ and using the material properties listed for steel in Table 1, the percentage error in measured axial and transverse strains will be 0 percent and -9.8 percent, respectively. Now consider a biaxial-strain-gage rosette mounted on an off-axis graphite/epoxy tensile specimen, as shown in Fig. 4 (again with $\beta = 0$). Both the axial and transverse strains measured will be in error due to transverse sensitivity. The magnitude of the error will vary as a function of the fiber angle Θ. Assuming $K = 0.03$ and using the material properties listed in Table 1, it can be shown[11] that the actual and measured axial and transverse strains will vary as indicated in Figs. 7 and 8, respectively. The error in axial-strain measurement indicated in Fig. 7 is very small, and for all fiber angles the percentage error is less than 1 percent. However, the errors in transverse-strain measurement shown in Fig. 8(a) are quite large; the maximum measurement error occurs for a fiber angle of 90°, where the actual and measured strains are -113μm/m and +113μm/m, respectively. Percentage errors are shown in Fig. 8(b), and range from -9 to -200 percent! Obviously these strain measurements must be corrected for transverse-sensitivity effects.

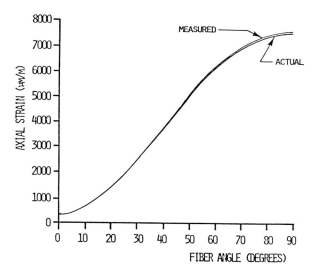

Fig. 7—Actual and measured axial strain (prior to correction for transverse sensitivity effects) as a function of fiber angle ($\sigma_x = 60$ MPa)

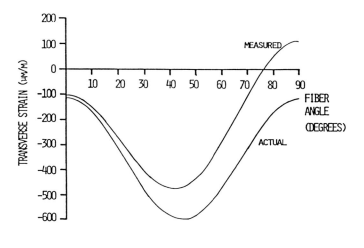

Fig. 8 (a)—Actual and measured transverse strain (prior to correction for transverse sensitivity effects) as a function of fiber angle ($\sigma_x = 60$ MPa)

This example illustrates that transverse-sensitivity effects must *always* be considered when dealing with strain gage measurements obtained for composites. In some instances transverse-sensitivity effects are negligibly small and can be safely ignored. In other cases, gross measurement errors occur which must be accounted for.

Strain Measurement Near a Free Edge

A final caution regarding strain measurement near a 'free edge' is in order. The term 'free edge' refers to a boundary of a composite structure which is not subjected to any external loading. Typical examples include the two unloaded sides of a uniaxial-tensile specimen or a cutout in a composite panel. Since most composite panels are relatively thin plate-like structures, it is appropriate to analyze composites using thin-plate theory. The combination of orthotropic elasticity and thin-plate theory results in so-called 'classical lamination theory' (CLT)[7,8]. CLT is based upon the plane-stress assumption and the Kirchhoff hypothesis, i.e., a line which is initially straight and perpendicular to the midplane of the composite plate is assumed to remain straight and perpendicular to the midplane after deformation. Together these assumptions imply that out-of-plane normal and shear stresses, usually denoted σ_z, τ_{xz}, and τ_{yz}, are all zero. These assumptions are well satisfied at regions removed from a free edge, and consequently surface strain measurements (which might be obtained using strain gages, for example) can be used to infer subsurface strains in these regions. However, near a free

Fig. 8(b)—Percentage error in measured transverse strain (prior to correction for transverse sensitivity effects) as a function of fiber angle

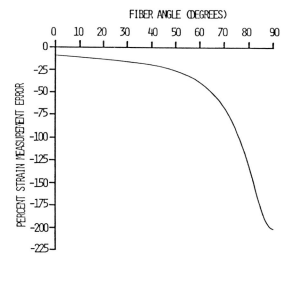

edge *neither* the plane-stress assumption *nor* the Kirchhoff hypothesis is valid. That is, near a free edge, significant out-of-plane stresses (σ_z, τ_{xz}, τ_{yz}) and out-of-plane strains (ε_z, ε_{xz}, ε_{yz}) are all induced[12,13]. Near a free-edge, surface strain measurements may not be related to subsurface strains. Therefore caution must be exercised when applying strain gages near a free edge—when attempting to measure strain concentrations near a cutout in a composite panel, for example.

References

1. Dally, J.W. and Riley, W.F., *Experimental Stress Analysis*, 2nd Ed., McGraw-Hill Book Co., New York, NY (1978).

2. Perry, C.C. and Lissner, H.R., *The Strain Gage Primer*, 2nd Ed., McGraw-Hill Book Co., New York, NY (1962).

3. "Optimizing Strain Gage Excitation Levels," M-M Tech Note TN-502, Measurements Group, Inc., Micro-Measurements Div., Raleigh, NC.

4. "Standard Test Method for Performance Characteristics of Bonded Resistance Strain Gages," ASTM Standard E251-67.

5. "Errors Due to Transverse Sensitivity in Strain Gages," M-M Tech Note TN-509, Measurements Group, Inc., Micro-Measurements Div., Raleigh, NC.

6. "Errors Due to Misalignment of Strain Gages," M-M Tech Note TN-511, Measurements Group, Inc., Micro-Measurements Div., Raleigh, NC.

7. Jones, R.M., *Mechanics of Composite Materials*, McGraw-Hill Book Co., New York, NY (1975).

8. Halpin, J.C., *Primer on Composite Materials: Analysis*, Technomic Publishing Co., Inc., Lancaster, PA (1984).

9. Greszczuk, L.B., "Effect of Material Orthotropy on the Directions of Principal Stresses and Strains," *Orientation Effects in the Mechanical Behavior of Anisotropic Structural Materials*, ASTM STP 405, 1-13 (1965).

10. Tuttle, M.E., and Brinson, H.F., "Resistance Foil Strain Gage Technology as Applied to Composite Materials", EXPERIMENTAL MECHANICS, **24** (1), 54-65 (1984).

11. Tuttle, M.E., "Error in Strain Measurements Obtained Using Strain Gages on Composites", Proc. 1985 SEM Fall Mtg, Nov. 17-20, 170-179 (1985).

12. Pipes, R.B. and Pagano, N.J., "Interlaminar Stresses in Composite Laminates Under Uniform Axial Extension", J. COMP. MAT., **4**, 538-548 (1970).

13. Pagano, N.J. and Pipes, R.B., "The Influence of Stacking Sequence on Laminate Strength", J. COMP. MAT., **5**, 50-57 (1971).

Section IIIB

Strain Gages on Composites—Gage-Selection Criteria

by R. Slaminko

Selection of the proper strain gage is one of the first steps toward a successful test. The test material, the test environment, and the goals of the strain-gage survey must all be considered when choosing a strain gage, as well as the total cost of the gage installation.

The following parameters of a strain gage are generally user-specified and must be considered: strain-gage material, including the gage alloy and the backing material, strain-gage sensing length, gage width, gage resistance, gage pattern, and self-temperature compensation number. The primary controlling factors to be considered are generally: specimen geometry and surface contours, thermal expansion and thermal conductivity properties of the test material, test environment, test duration, in both time and number of load cycles, expected strain levels and strain gradients, required levels of accuracy and stability, and simplicity and cost of the system.

Gage selection for composites is not generally more difficult than for other materials, but the following characteristics of composites should be kept in mind: (1) The strain field of a composite exhibits local strain variations due to both the nonhomogeniety of the material and the surface texture created by the scrim cloth. (2) Composites are poor thermal conductors. (3) The effective coefficient of thermal expansion of a composite varies both with direction and with the previous load and thermal history of the specimen. (4) Principal strains and stresses may not be coaxial due to the anisotropic nature of composites. (5) Maximum strain levels are usually less than three percent. (6) Composites are not commonly utilized in hostile environments which would damage most modern strain gages.

Selection of Strain-Gage Material

A strain gage may be thought of as consisting of two constituent materials: the alloy of the sensing grid and the backing, or carrier, material. The most common gage alloys for use on composites are constantan and Karma. Both materials exhibit good sensitivity, stability, and fatigue life, and can be fabricated to present a variety of self-temperature-compensation values (although this property is not of great significance for composites, as will be discussed later). Karma, however, is more stable above $+65°$C ($+150°$ F) than constantan, and is usable over the range of $-270°$ C to $+290°$ C ($-452°$ F to $+550°$

F). Karma's stability makes it superior for long-term (months or years) accurate strain measurements. Karma, however, is more difficult to solder to than constantan, so it may be advisable to purchase gages with preattached leadwires (to prevent the possibility of heat-induced damage to the composite, it may be advisable to either buy prewired gages or to attach the leadwires prior to installing the gage regardless of the alloy chosen).

The backing material serves to maintain the geometry of the gage during handling and to electrically insulate the gage element from the test article. The criteria for selecting a carrier are no more stringent for composites than for conventional materials and may be less so, as composites rarely exceed three percent maximum strain and are poor electrical conductors (so insulation is not critical). The most important considerations for selecting a backing material are contourability, temperature stability and maximum elongation. Polyimide-type materials are best in terms of contourability and maximum elongation, while glass-fiber-reinforced phenolics offer the best temperature stability.

Gage Size and Resistance Considerations

The size of a strain gage is important for strain averaging and power density. Composite materials are more sensitive to these effects than other materials because of their local strain variations and their poor thermal conductivity.

A strain gage indicates the average strain under the grid area. In areas of high strain gradient, either parallel or perpendicular to the sensing axis, this can cause the gage to report a strain level considerably less than the peak strain, which is usually the desired quantity. Use of shorter, narrower gages will minimize this averaging effect, but will magnify any errors due to gage mislocation.

Short strain gages (less than 3 mm in length) have several other disadvantages as well, including degraded stability, endurance to cyclic strain, maximum elongation, and handling characteristics. They are also generally more expensive and have less options available than larger gages. On composites, two other disadvantages of short gages become important. First, many composites present a 'weave' texture due to the scrim cloth which can cause local strain magnitude variations of as much as 10 percent;

it is generally preferable to measure the average strain in such areas, not the local peaks and valleys. Second, smaller gages present a higher power density than larger gages of the same resistance for a given excitation level.

Power density is an important consideration for composites due to their poor thermal-conductivity properties. The performance of a strain gage mounted on a poor thermal conductor is degraded in a number of significant ways, including increased drift, hysteresis, and creep effects as well as loss of self-temperature-compensation. Additionally, excess heat generated by a strain gage may introduce local stresses or material property changes in the underlying composite as well as increase the creep rate. The effective power density of a gage is dependent on three factors: the area of the grid (larger is better), the resistance of the gage (higher is better), and the excitation level (lower is better, as long as the level is sufficiently high to avoid excess noise and decreased sensitivity). Acceptable power-density levels for strain gages on composites are generally in the range of 0.31 kW/m² to 1.20 kW/m² (0.2 W/in.² to 0.77 W/in.²).

With the above considerations in mind, the following are rule-of-thumb guidelines for selecting gage size, resistance, and excitation levels:

Size: 3 mm (0.125 in.) or larger gages
Resistance: 350 Ω or highe· and
Excitation level: 3 v or less.

Smaller gages may be used, if necessary, by reducing the excitation level.

Gage-Pattern Selection

The 'pattern' of a strain gage refers to whether a gage is an axial, T-rosette, or three-legged rosette, and also to whether the gage elements are stacked or displaced (uniplanar). Axial gages are of only limited use on composite materials. The axes of principal stress and strain may not coincide on these materials because of their anistropic nature, even if a uniaxial stress state can be assumed. Free boundaries are the only safe locations for axial gages on composites. T-rosettes, consisting of two perpendicular elements, are also of limited use for these same reasons. Three element rosettes, whether 45-deg. rectangular or 60-deg. delta configuration, are the best choice for general analysis of composite structures. This type of gage is the only one that guarantees adequate information to solve the general strain state of a composite material, and can be installed with virtually any orientation without loss of accuracy.

The choice between stacked or displaced rosettes will depend on several factors: presence of a strain gradient normal to the test surface, severity of the strain gradients in the plane of the test surface, thermal considerations, and installation space available.

Stacked rosettes require less installation space and are also less sensitive to in-plane gradients, but are inferior in terms of stability and accuracy for static measurements due to increased heat transfer to the test specimen. They are also less accurate than uniplanar rosettes in the presence of strain gradients normal to the test surface, such as would occur in bending.

Self-Temperature-Compensation Number

Self-temperature-compensated (S-T-C) gages are constructed of a material which has been processed to produce minimum output due to temperature variations when mounted to a material whose coefficient of thermal expansion matches the S-T-C number of the gage. This type of gage is not well suited for achieving adequate temperature compensation on composites, however, mainly because the coefficient of thermal expansion of a composite varies with direction, stacking sequence, past thermal and mechanical history, moisture content, and degree of flaws and damage (e.g., matrix cracks, delaminations, voids). This makes it virtually impossible to select a gage with an appropriate S-T-C number. Adequate temperature compensation can only be reliably achieved via either dummy gages or precalibration. These techniques are covered in greater detail elsewhere in this manual. For simplicity's sake, it would probably be best to choose gages with an S-T-C number of 0.

Summary

Selection of the proper strain gage for use on a composite material requires consideration of many aspects of the test material, the test environment, and the gage itself. The following guidelines will, in most cases, provide a good starting point for gage selection:

Gage alloy:	Karma or constantan
Backing:	polyimide or phenolic
Resistance:	350 Ω or greater
Size:	3 mm (0.125 in.) or larger
Pattern:	Three-element rosette, except at free boundaries Displaced elements, unless space restrictions or in-plane gradients necessitate the use of stacked elements
S-T-C number:	unimportant, but for simplicity, should be near 0

It may also be wise to select gages with preattached leadwires to prevent thermal damage to the composite during soldering.

Bibliography

Chamis, C.C. and Sinclair, J.H., "Ten-Degree Off-Axis Test for Shear Properties in Fiber Composites," EXPERIMENTAL MECHANICS, 17, 339-346 (1977).

Measurements Group Inc., "Optimizing Strain Gage Excitation Levels," Measurements Group Tech Note TN-502 (1979).

Measurements Group Inc., "Strain Gage Selection-Criteria, Procedures, Recommendations," Measurements Group Tech Note TN-505 (1983).

Perry, C.C., "The Resistance Strain Gage Revisited," EX-PERIMENTAL MECHANICS, 24 (4), 286-299 (Dec. 1984).

Perry, C..C. and Lissner, H.R., The Strain Gage Primer, McGraw-Hill Book Co., (1962).

SEM, Handbook on Experimental Mechanics, Prentice-Hall Inc., 41-78, 814-885 (1987).

Tuttle, M.E., and Brinson, H.F., "Resistance-Foil Strain Gage Technology as Applied to Composite Materials," EXPERIMENTAL MECHANICS, 24, (1), 54-65 (March 1984).

Whitney, J.M., Daniel, I.M., and Pipes, R.B., "Experimental Mechanics of Fiber Reinforced Composite Materials," SESA, (1982).

Section IIIC

Strain Gages on Composites Temperature Compensation

by R. Slaminko

Introduction

When a strain gage is mounted on a test article which is subjected to both mechanical loads and temperature variations, the output of that gage will represent a combination of the strains induced by both effects. For most experimental-mechanics work, it is desirable to consider only that portion of the output which is representative of the stresses applied to the test article. It is therefore necessary to somehow separate this desired output from the unwanted effects of the temperature variations. This separation is achieved by a process known as 'temperature compensation'.

The portion of the output of a strain gage which results from temperature variations is called apparent strain. It occurs for two basic reasons: the electrical resistance of a strain gage changes with temperature, and the coefficient of thermal expansion (amount of expansion/contraction per degree temperature change) of the gage is probably somewhat different than that of the material to which it is attached. Apparent strain can be one of the most serious sources of error in strain-gage measurements, hence the need for temperature compensation.

Fig. 1—Variation in CTE of unidirectional graphite-epoxy composite with direction. Direction measured from fiber direction

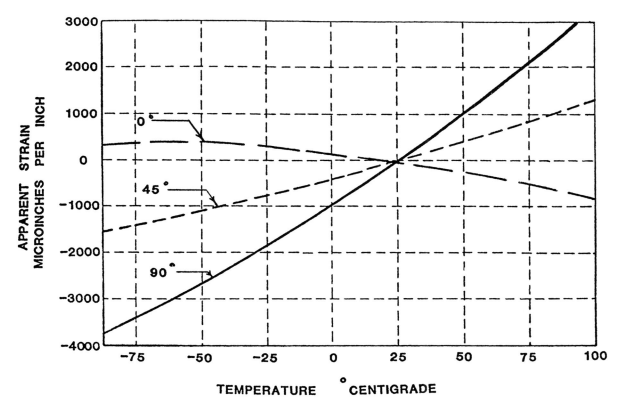

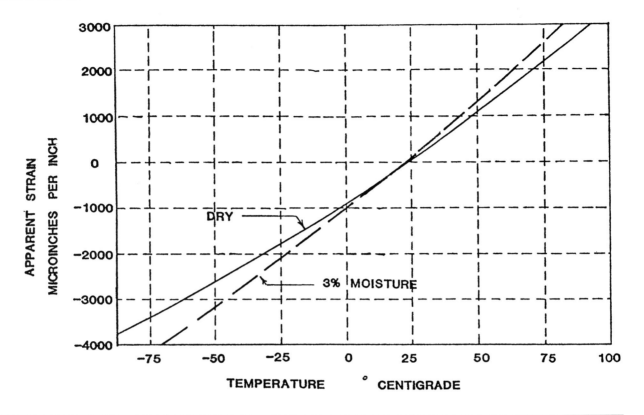

Temperature compensation is more complex when dealing with composite materials than with conventional homogeneous isotropic structures. The reasons for this increased complexity can be summarized as follows. (1) The coefficient of thermal expansion (CTE) of a single composite ply is a highly anisotropic property, in general being very low in the fiber direction and quite high perpendicular to the fibers. This effect is illustrated in Fig. 1. (2) The CTE of a given material varies somewhat from manufacturer to manufacturer and may even differ slightly between lots from the same manufacturer. (3) The CTE of a composite layup will vary not only with the makeup of the individual plies, but also with the stacking sequence. (4) The past mechanical and thermal history of the test article affects the CTE; the rate of temperature change can also have an effect. (5) Voids, delaminations, matrix cracks, and moisture content all will alter the CTE to some extent. Fig. 2 illustrates the effect of moisture content.

Despite these difficulties, correction for the apparent strain due to temperature variation can be achieved for composite materials. Three methods for achieving temperature compensation are: self-compensated gages, dummy gages, or precalibration.

Each of these methods has certain advantages and limitations and will therefore be discussed separately.

Self-Temperature Compensation

Self-temperature-compensated (S-T-C) gages are generally the quickest, least expensive, and most widely used (on conventional materials) method for achieving apparent-strain correction. The primary attraction of the technique is that it requires neither dummy gages nor a pre-test calibration cycle. Data produced by an S-T-C gage are already corrected for the apparent strain due to temperature variations, and no additional data processing is required. This is accomplished by selecting a strain gage alloy (generally constantan or Karma) which has been processed to present a nearly zero output in response to temperature variations when mounted to a material whose coeffecient of thermal expansion matches the S-T-C number of the gage. Figure 3 shows apparent-strain curves generated by typical commercially available constantan and Karma gages mounted on steel. As shown, reasonably accurate apparent-strain compensation is achieved over the range of approximately -20 to +205°C (0 to 400°F).

The use of S-T-C gages on composite materials is more problematic. The coefficient of thermal expansion for these materials varies somewhat between manufacturers and even between lots from the same manufacturer, as well as between stacking sequences. These considerations make it virtually impossible to either maintain a supply of gages with the proper S-T-C, or to even know the effective coefficient of thermal expansion of the test material without prior testing.

Even if the above problems could be solved, S-T-C gages are still not the best choice for composites because of the directional sensitivity of the coefficient of thermal expansion of a composite. This would force the use of gages with different S-T-C values in every desired measurement direction. Additionally, this variation in coefficient of thermal expansion can introduce transverse-sensitivity errors into the data produced by a strain gage.

Dummy-Gage Compensation

The second, and more common in the case of composites, method of thermal compensation is the use of dummy gages. In theory, the dummy gage (which should be of an identical configuration as the active gage) is subjected to the same thermal environment as the active gage but remains unaffected by the mechanical forces applied to the structure on which the active gage is mounted. The leadwires for the two gages must be of equal length and subjected to identical environments also. If these conditions are met and the active and dummy gages are installed in adjacent arms of a Wheatstone bridge circuit as illustrated in Fig. 4, automatic thermal compensation will occur.

Dummy-gage compensation requires somewhat more care when used on composites than is necessary with isotropic materials. The composite on which the dummy gage is mounted must be as identical as possible to that which holds the active gage. This includes basic materials, stacking sequence, thermal and mechanical load history, and moisture content. Additionally, the dummy gage must be installed with the same orientation (relative to the major material axis) as the active gage; relative mismatches of as little as four deg. have been shown to result in errors of up to two microstrain per degree Celsius. The significance

Fig. 3—Typical apparent-strain curves from self-compensated constantan and Karma gages mounted on steel

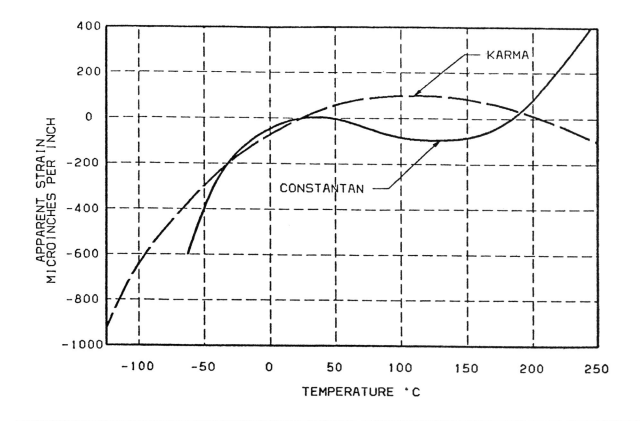

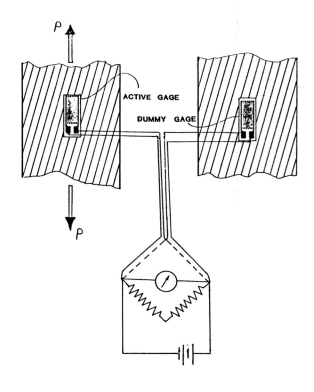

Fig. 4—Dummy-gage compensation system

of this error will of course depend on the temperature range of the test and the magnitude of the mechanically induced strain. Because the coefficient of thermal expansion of composites varies somewhat with mechanical and thermal history, it may be necessary to duplicate the history of the active specimen onto the dummy to maintain accurate compensation.

The primary drawback to dummy-gage compensation is the necessity of a second, unloaded test article on which to mount the dummy gage. In most cases, however, a simple specimen fabricated from the same material as the test article will suffice for holding the dummy gage. The dummy gages may even be mounted to the same structure as the active gages if it can be guaranteed that the portion of that structure with the dummy gage will be completely unaffected by the mechanical loading. If the dummy-gage system is implemented with appropriate care, it can be used to provide very accurate thermal compensation over a wide temperature range.

Precalibration

A third method of achieving compensation is to precalibrate the strain gages. This may be accomplished by thermally cycling the test structure throughout the anticipated temperature range of the test, without applying mechanical loads, and generating an apparent-strain curve. Then, when the actual test is conducted, the apparent strain can be subtracted from the gage output to yield the

mechanically induced strain. The primary advantages of this method are: (1) very accurate thermal compensation is possible over the entire temperature range; and (2) the active gage is used to generate the apparent-strain curve; therefore, no errors due to mismatch of gage type or orientation, or with material inconsistencies can occur. The major disadvantages are: (1) a separate calibration run is required prior to the actual test; (2) an accurate measure of the temperature must be recorded with each data scan; and (3) the data must be postprocessed to achieve compensation—although it is possible to automate this process.

It should also be emphasized that, since the effective coefficient of thermal expansion of a composite varies with moisture content and load history, it is important to ensure that the calibration run is performed with the specimen in the same condition as will exist for the actual test. This may require a preliminary heating of the specimen to drive off excess internal moisture.

Conclusion

Three techniques for correcting strain gage data obtained from composite materials for temperature-induced apparent strain have been discussed: self-temperature-compensated gages, dummy gages, and precalibration. S-T-C gages are generally unsuited for use on composites due to the anisotropic nature of the coefficient of thermal expansion of those materials. Dummy gages can provide excellent thermal compensation if it can be assured that the dummy gage and the material to which it is mounted match the active gage and actual test material exactly. Precalibration is another viable option for achieving temperature compensation, but requires additional data processing.

Bibliography

Cairns, D.S. and Adams, D.F., "Moisture and Thermal Expansion of Composite Materials", Univ. of Wyoming Dept. of Mech. Eng. Rept. UWME-DR-101-104-1 (1981).

Chamis, C.C., "Simplified Composite Micromechanics Equations for Hygral, Thermal, and Mechanical Properties", SAMPE Quarterly (April 1984).

Delasi, R. and Whiteside, J.B., "Effect of Moisture on Epoxy Resins and Composites", ASTM STP-658 (1978).

Measurements Group Inc., "Measurement of Thermal Expansion Coefficient Using Strain Gages", Measurements Group Tech Note TN-513 (1986).

Measurements Group Inc., "Temperature-Induced Apparent Strain and Gage Factor Variation in Strain Gages", Measurements Group Tech Note TN-504 (1983).

Perry, C.C., "The Resistance Strain Gage Revisited", EXPERIMENTAL MECHANICS, 24 (4), 286-299 (Dec. 1984).

SEM, Handbook on Experimental Mechanics, Prentice-Hall Inc., 55-57, 842-843 (1987).

Tuttle, M.E. and Brinson, H.F., "Resistance-Foil Strain Gage Technology as Applied to Composite Materials", EXPERIMENTAL MECHANICS, 24 (1), 54-65 (March 1984).

Whitney, J.M., Daniel, I.M., and Pipes, R.B., "Experimental Mechanics of Fiber Reinforced Composite Materials", SESA (1982).

Section IIID

Strain-Gage Reinforcement Effects on Low-Modulus Materials

by C.C. Perry

It has been demonstrated, both experimentally and analytically, that the stiffness of a strain gage can produce a significant reinforcing effect when the gage is installed on a material with a low elastic modulus — say, 1×10^6 psi (7 GPa) or less. In tests of strain gages on plastics, for example, errors ranging from -10 percent to -30 percent and greater have been reported.[1,2] In principle, the errors could be eliminated by calibrating the gages for their *effective* gage factors when mounted on plastics. Considering, however, the wide range and variability of mechanical properties found in plastics, there is presently no practical means for the gage manufacturer to supply such information. Nor is it ordinarily feasible for most stress-analysis laboratories to undertake the difficult and exacting task of gage-factor calibration, even for a single gage type on a specific plastic material. For purposes of routine experimental stress analysis of products made from plastics (or other low-modulus materials), this paper offers a relatively simple means of compensation for strain-gage-reinforcement effects.

The reinforcement by the gage can be characterized as either *local* or *global*, or some combination of the two. When the cross section of the test member at the gage location is great enough (for a particular elastic modulus) that the contribution of the gage to bearing the applied bending moments or in-plane loads is negligible, the reinforcement is defined here as local. Under these circumstances, the far-field strains surrounding the gaged section are essentially unaffected by the gage installation, and perturbation of the strain field occurs only in the area immediately under and around the gage. When, however, the cross section at the gage site is sufficiently thin and narrow that the gage stiffness represents a sensible fraction of the section stiffness, the strain magnitudes throughout the section are altered by the presence of the gage, and the reinforcement is described as global. Intermediate sectional properties can result in any degree of mixed local and global reinforcement.

For many load-bearing applications of plastics in industrial and consumer products, the sectional properties in regions where strain measurements must be made are such that the reinforcement caused by a strain gage can be considered primarily local.

The present study is restricted to this class of applications, and is further limited to plastics which are effectively isotropic in their elastic properties. The anisotropy associated with directionally reinforced plastics introduces additional constraints, and is the subject of a separate analysis.

When significant reinforcement is present, the strain transmitted to the gage grid differs from the unperturbed strain at the gage site, and the gage output is correspondingly in error unless the effect is compensated or corrected for. From the following general expression for gage output, it appears necessary to consider not only the reinforcement along the primary sensing axis of the gage, but that in the transverse direction as well:

$$\frac{\Delta R}{R} = F_a \varepsilon_a + F_t \varepsilon_t \qquad (1)$$

where

$\varepsilon_a, \varepsilon_t$ = axial (longitudinal) and transverse strains sensed by the gage

F_a, F_t = axial and transverse gage factors

Consider first the gage response under the standard conditions employed by the manufacturer in calibrating for gage factor. For this purpose, the gage is installed on a steel beam of generous cross section so that reinforcement is negligible. Since the surface of the beam is in a state of uniaxial stress, and the gage is aligned along the beam axis,

$$\varepsilon_t = -\nu_o \varepsilon_a \qquad (2)$$

where ν_o = Poisson's ratio of steel calibration beam, usually taken as 0.285. Thus,

$$\frac{\Delta R}{R} = \varepsilon_a (F_a - \nu_o F_t) \qquad (3)$$

The gage factor specified by the manufacturer is then defined as follows:

$$GF = \frac{\Delta R/R}{\varepsilon_a} = F_a - \nu_o F_t \qquad (4)$$

It is obvious from the foregoing that the axial strain inferred from the unit resistance change and manufacturer's gage factor is correct only for the calibration conditions; i.e., when $\varepsilon_t = -\nu_o\varepsilon_a$. In any other strain field with a different ratio of axial to transverse strains, the calculated strain along the gage axis is in error. To permit the gage user to correct for this error, the manufacturer also measures the transverse response of the gage. The standard procedure for doing this is to use a special test fixture which produces a uniaxial strain field. A pair of identical gages is installed on the fixture, with one gage aligned perpendicular to the strain field, and the second gage parallel to it. With this arrangement, the axial strain ε_a is zero for the first gage, and the transverse strain ε_t is zero for the second. Thus, after applying eq (1) to each gage, the ratio of the first to second gage outputs is F_t/F_a. This parameter is customarily referred to as the 'transverse sensitivity' of the gage, defined by

$$K_t = \frac{F_t}{F_a} \qquad (5)$$

Equation (4) then becomes

$$GF = \frac{\Delta R/R}{\varepsilon_a} = F_a(1 - \nu_o K_t) \qquad (6)$$

Thus, both the axial and transverse responses must, in general, be considered to accurately determine the strain along the gage axis. Special relationships based on eqs (1) and (6), and implicitly assuming freedom from reinforcement effects, have been developed to permit the gage user to correct for the transverse sensitivity of single-element and rosette gages in any strain field.[5]

It appears reasonable to assume that biaxial reinforcement effects can be adequately modeled by introducing two additional variables into eq (1) as follows:

$$\frac{\Delta R}{R} = F_a\lambda_a\varepsilon_a + F_t\lambda_t\varepsilon_t \qquad (7)$$

where λ_a, λ_t = strain-transmission coefficients. The coefficients λ_a and λ_t represent, respectively, the fractions of the surface strains ε_a and ε_t that are transmitted to the gage grid under reinforcement conditions. Alternatively, the products $F_a\lambda_a$ and $F_t\lambda_t$ can be looked upon as the *effective* axial and transverse gage factors applicable to the same conditions. It is assumed that λ_a and λ_t are independent of the strain level, and are functions only of gage proportions and the ratio E_p/E_g, where E_p is the elastic modulus of the plastic or other test material and E_g the 'equivalent modulus' of the gage. When the gage is installed on a metal test member where reinforcement is negligible, both coefficients must closely approach unity in order that eq (7) revert to eq (1). Judging from the experimental

data for gages applied to plastics, the coefficients tend to decrease as some function of the ratio E_p/E_g, reflecting a reduction in strain transmitted to the grid as the elastic modulus of the test material becomes lower. Although both λ_a and λ_t may be characterized by the same function, they are treated here as separate coefficients for the sake of generality.

Pending the development of more rigorous, standardized calibration procedures, the method described here should permit considerably more accurate determination of stresses than can be achieved by the common practice of ignoring reinforcement effects altogether. The technique consists of measuring the *apparent* elastic properties of the plastic, using the identical type of strain gage that will be employed in subsequent stress-analysis measurements on actual test parts. It can be shown that errors in indicated strain due to gage reinforcement effects (and transverse sensitivity) are then canceled by the inverse errors in the elastic properties when measured strains are converted to stresses via Hooke's law. An important practical advantage of the method arises from the fact that both the indicated strains and apparent elastic properties can be measured quite easily, and with relatively high accuracy. The general principle underlying this approach was described as long ago as 1968 by Meyer[6]. Although never widely applied in the past, the technique may now represent the most effective practical means for reasonably accurate experimental stress analysis of load-bearing plastic parts.

The method can be implemented by first fabricating a calibration specimen from the intended test material. Calibration is most easily accomplished with a simple tensile or bending specimen, designed for uniform uniaxial surface stress. Specimen section dimensions should be great enough to assure only local reinforcement by strain gages. Gages are then installed on the specimen to indicate longitudinal and transverse strains under load.*

With the gage-factor setting of the instrumentation at the manufacturer's value GF, the strain indicated by the longitudinal gage is:

$$\hat{\varepsilon_1} = F_a(\lambda_a\varepsilon_a + K_t\lambda_t\varepsilon_t)/GF \qquad (8)$$

where

$\hat{\varepsilon_1}$ = indicated strain
ε_a = longitudinal surface strain
ε_t = transverse surface strain

---

* Good strain-gage practices, and good measurement practices generally, are assumed throughout this procedure. Examples include: use of back-to-back gages on a tensile specimen to eliminate bending effects, proper gage bonding and installation techniques, low gage excitation voltage to avoid sel-heating effects, isothermal testing, etc.

Since, for uniaxial stress, $\varepsilon_t = -\nu_p\varepsilon_a$, where ν_p is the Poisson's ratio of the test material,

$$\hat{\varepsilon}_1 = F_a\varepsilon_a(\lambda_a - \nu_pK_t\lambda_t)/GF \qquad (9)$$

Correspondingly, for the transverse gage,

$$\hat{\varepsilon}_2 = F_a\varepsilon_a(-\nu_p\lambda_a + K_t\lambda_t)/GF \qquad (10)$$

The apparent Poisson's ratio of the material is then:

$$\nu_p^* = \frac{-\hat{\varepsilon}_2}{\hat{\varepsilon}_1} = \frac{\nu_p\lambda_a - K_t\lambda_t}{\lambda_a - \nu_pK_t\lambda_t} \qquad (11)$$

And the apparent elastic modulus becomes:

$$E_p^* = \frac{\sigma_1}{\hat{\varepsilon}_1} \quad \frac{\sigma_1 \bullet GF}{F_a\varepsilon_a(\lambda_a - \nu_pK_t\lambda_t)}$$

where $\sigma_1 = $ applied calibration stress. Noting that $\sigma_1/\varepsilon_a = E_p$, the actual elastic modulus of the test material, and that $GF = F_a(1 - \nu_oK_t)$,

$$E_p^* = \frac{E_p(1 - \nu_oK_t)}{\lambda_a - \nu_pK_t\lambda_t} \qquad (12)$$

Subsequently, in the stress analysis of a part made from the same material, the indicated strains in any two perpendicular directions will be

$$\hat{\varepsilon}_x = F_a(\lambda_a\varepsilon_x + K_t\lambda_t\varepsilon_y)/GF \qquad (13)$$

$$\hat{\varepsilon}_y = F_a(\lambda_a\varepsilon_y + K_t\lambda_t\varepsilon_x)/GF \qquad (14)$$

Assuming linear-elastic behavior of the test material, and writing the biaxial Hooke's law in terms of indicated strains and apparent elastic properties,

$$\sigma_x = \frac{E_p^*}{1 - \nu_p^{*2}}(\varepsilon_x + \nu_p^*\varepsilon_y) \qquad (15)$$

Substituting into eq (15) from eqs (11), (12), (13), and (14), and reducing, demonstrates that

$$\hat{\sigma}_x = \frac{E_p}{1 - \nu_p^2}(\varepsilon_x + \nu_p\varepsilon_y) = \sigma_x$$

Thus, the reinforcement effects and transverse-sensitivity errors in the indicated strains are canceled by those in the apparent elastic properties, and the indicated stress is equal to the actual stress in the test member. The same is true, of course, for the stress in the Y direction.

The foregoing result could have been anticipated in the light of earlier work on the subject of transverse-sensitivity errors and compensation/correction methods.[6-8] It is generally recognized that the indicated strains from gages with biaxial sensitivity

transform in the usual manner according to Mohr's circle, paralleling the physical strains in the test surface. Based on the same consideration, Gu recently proposed the technique described here as a means of intrinsic compensation for transverse-sensitivity errors.[9] As demonstrated by Meyer, and confirmed here, however, the method is very broadly applicable, and can be used to compensate for any constant linear errors in the variables which enter into Hooke's law.

Although the example used in this report to illustrate simultaneous cancellation of transverse-sensitivity and reinforcement errors referred to the use of only two perpendicular gages on the test object, the procedure is also applicable, in principle, to three-element rosettes as well. It is necessary in this case, of course, that the identical type of rosette be used, both on the calibration specimen for measuring the apparent elastic properties, and on the test object to measure the indicated working strains. To the degree that the gage factors, transverse sensitivities and strain-transmission coefficients may differ somewhat from element to element in the rosette, error cancellation will be accordingly less than complete. The residual error, however, should normally be much smaller than that due to uncompensated reinforcement effects.

As a result of the earlier assumption that the strain-transmission coefficients are independent of strain level, the coefficients disappear during data reduction, and it is never necessary (for stress-analysis purposes) to know either their functional forms or actual magnitudes. Equations (11) and (12) offer, however, the possibility of evaluating the coefficients in specific cases when the true material properties are known. Assume, for instance, that the true elastic properties of the test material have been determined accurately from independent non-perturbing measurements (by optical methods, for example). Then, after the apparent properties are measured with a particular type of strain gage installed on the material, the only remaining unknowns in eqs (11) and (12) are the strain-transmission coefficients. In principle, at least, λ_a and λ_t can be evaluated by solving these equations simultaneously.

$$\lambda_a = \frac{E_p(1 - \nu_oK_t)}{E_p^*(1 - \nu_p^2)} \bullet (1 - \nu_p\nu_p^*) \qquad (16)$$

$$\lambda_t = \frac{E_p(1 - \nu_oK_t)}{E_p^*(1 - \nu_p^2)} \bullet \frac{(\nu_p - \nu_p^*)}{K_t} \qquad (17)$$

Summarizing the procedure described here, the apparent elastic properties of the low-modulus test material are first measured with the same (identical) type of strain gage to be used in experimental stress analysis of parts made from the material. Subsequently, in data reduction to determine working stresses, the indicated strains measured on the test object are substituted into the expression for Hooke's law, along with the apparent elastic properties. The errors due to local strain-gage reinforcement and transverse sensitivity are then canceled in the data-reduction process, yielding the actual stresses in the loaded test object.

It can be expected, of course, that some of the existing published data on the elastic properties of commercial plastics were measured with strain gages initially. In such cases, the data may be already in error to a degree, depending on the extent of the reinforcement effect present in the measurements. When these properties are employed in Hooke's law, along with strain data which also include reinforcement errors, at least partial compensation of the errors should occur by default. However, strain-gage stiffness is known to vary significantly with the gage type. It would seem, therefore, that a substantial improvement in the overall accuracy of experimental stress analysis on low-modulus materials can be realized by both calibrating the elastic properties and measuring the working strains with the identical gage type.

References

1. McCalvey, L.F., "Strain Measurements on Low Modulus Materials," presented at the BSSM Conf., University of Surrey, U.K. (Sept. 1982).
2. White, R.N., "Model Study of the Failure of a Steel Bin Structure," presented at the ASCE/SESA Exchange Session on Physical Modeling of Shell and Space Structures, ASCE Annual Convention, New Orleans, LA (Oct. 1982).
3. Stehlin, P., "Strain Distribution In and Around Strain Gages," J. Strain Anal., 7 (3), 228-235 (1972).
4. Beatty, M.F. and S.W. Chewning, "Numerical Analysis of the Reinforcement Effect of a Strain Gage Applied to a Soft Material," Inter. J. Eng. Science, 17, 907-915 (1979).
5. Measurements Group, Inc., "Errors Due to Transverse Sensitivity in Strain Gages," Tech. Note TN-509 (1982).
6. Meyer, M.L., "On a General Method of Compensation in Strain Gauge Work," Strain, 4 (1), 3-8 (Jan. 1968).
7. Meir, J.H. and Mehaffey, W.R., "Electronic Computing Apparatus for Rectangular and Equiangular Rosettes," Proc. SESA, II (1), 78-101 (1944).
8. Nasudevan, M., "Note on the Effect of Cross-Sensitivity in the Determination of Stress," Strain, 7 (2), 74-75 (April 1971).
9. Gu, W.-M., "A Simplified Method for Eliminating Error of Transverse Sensitivity of Strain Gage," EXPERIMENTAL MECHANICS, 22 (1), 16-18 (Jan. 1982).

Section IIIE

Strain-Gage Reinforcement Effects on Orthotropic Materials

by C.C. Perry

There is both experimental and analytical evidence that the stiffness of a strain gage can produce a significant reinforcement error when it is installed on a low-modulus material such as a plastic.[1-4] This raises the question of errors due to the same effect when strain measurements are made on some types of orthotropic materials (e.g., unidirectionally reinforced plastics) which are characterized by a low elastic modulus in at least the minor principal material direction. Actually, as indicated by the goniometric distribution of mechanical properties plotted in Fig. 1, the elastic modulus of such a material is typically low in most directions, and not far from that of the plastic matrix, except for an angular range of about ± 30 deg from the major principal material axis.

A method is described in Section IIID by which approximate compensation for reinforcement effects can be achieved when the material is isotropic in its elastic properties. The procedure involves calibration of the material for its apparent elastic properties (E, v), employing the identical type of strain gage intended for subsequent use in experimental-stress-analysis tests. Later, when indicated strains are converted to stresses with Hooke's law, based on the apparent elastic properties, the reinforcement errors (as well as those due to transverse sensitivity) are canceled in the data-reduction process. It is shown here that an extension of the same method can be applied, with certain restrictions, to some types of composite materials having directionally variable elastic properties.

For the purpose of this demonstration, a unidirectionally reinforced plastic has been selected as an example. The proposed method should be applicable, however, to other material types which conform to the same reinforcement model. An orthotropic material such as that considered here has four independent elastic constants, usually taken as E_1, E_2, v_{12}, and G_{12}. These represent the major and minor elastic moduli, the major Poisson's ratio, and the shear modulus, respectively. Since the normal-stress characteristics of the material (E_1, E_2, v_{12}) are commonly measured in separate tests from that used to determine the shear modulus; and since, with respect to the principal material axes, normal and shear responses are uncoupled, this method employs separate compensation of the normal and shear components.

With a strain gage installed on a metal surface, where reinforcement by the gage is negligible, the output of the gage can be expressed in the following general form:[5]

$$\frac{\Delta R}{R} = F_a \varepsilon_a + F_t \varepsilon_t \qquad (1)$$

where F_a, F_t = axial and transverse gage factors of strain gage, and $\varepsilon_a, \varepsilon_t$ = axial and transverse surface strains. When, on the other hand, the test material is low enough in elastic modulus that it is significantly

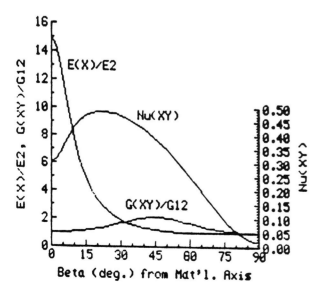

Fig. 1—Representative angular distribution of mechanical properties in an orthotropic material (graphite/epoxy). Graph produced by ORTHOLAM (lamina orthomechanics) program

reinforced by the gage, the strain transmitted to the gage grid differs from the unperturbed surface strain, and the gage output is altered correspondingly. The effect is modeled here, and in the preceding study for isotropic plastics, by introducing two additional variables into the expression for gage output:

$$\frac{\Delta R}{R} = F_a\lambda_a\varepsilon_a + F_t\lambda_t\varepsilon_t \qquad (2)$$

where λ_a, λ_t = strain-transmission coefficients.

The coefficients λ_a and λ_t represent, respectively, the fractions of the surface strains ε_a and ε_t that are transmitted to the gage grid under reinforcement conditions. Alternatively, the products $F_a\lambda_a$ and $F_t\lambda_t$ can be viewed as the *effective* axial and transverse gage factors applicable to the same conditions. It is assumed that λ_a and λ_t are independent of the strain level, and are functions only of gage proportions and the ratio E_i/E_g, where E_i is the relevant elastic modulus of the test material, and E_g the 'equivalent modulus' of the gage. In the case of a metal test member, with negligible reinforcement, both coefficients must closely approach unity in order that eq (2) effectively revert to eq (1). Judging from the experimental data for gages installed on plastics, the coefficients tend to decrease as some function of E_i/E_g, reflecting a reduction in strain transmitted to the gage grid as the elastic modulus of the test material becomes lower.[1,2] Although both λ_a and λ_t may be characterized by the same function, they are treated here as separate coefficients for the sake of generality.

Assume that a calibration specimen has been fabricated from a unidirectionally reinforced plastic as indicated in Fig. 2. Identical strain gages, aligned in the 1 and 2 directions, are installed on the specimen, which is then subjected to a uniaxial stress, σ_1. Although not drawn to scale in the illustration, the specimen cross section should be great enough to assure only *local* reinforcement effects by the gages. In other words, the gage stiffness should be small enough compared to the overall section stiffness that perturbation of the strain field is confined to the immediate vicinity of the gage.

Applying the model of eq (2) to this calibration specimen, the output of the gage aligned in the 1 direction can be expressed as:

$$\left(\frac{\Delta R}{R}\right)_1^1 = F_a\lambda_{1_a}\varepsilon_1^1 + F_t\lambda_{1_t}\varepsilon_2^1 \qquad (3)$$

where

$\left(\dfrac{\Delta R}{R}\right)_1^1$ = output of gage aligned in the 1 direction (subscript) due to uniaxial stress applied in the 1 direction (superscript)

$\lambda_{1\,a}, \lambda_{1\,t}$ = axial and transverse strain-transmission coefficients for a gage oriented in the 1 direction

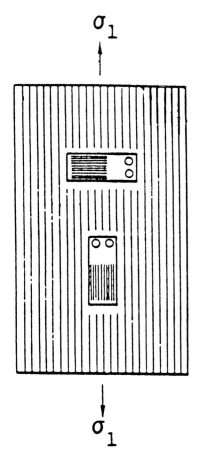

Fig. 2—Longitudinal specimen for evaluating apparent elastic properties E_1^* and v_{12}^*

$\varepsilon_1^1, \varepsilon_2^1$ = actual surface strains in the 1 and 2 directions (subscripts) due to uniaxial stress in the 1 direction (superscripts)

The relationship in eq (3) can be rendered more convenient for the present purposes if re-expressed in terms of the erroneous strain indicated by the gage under reinforcement conditions. Introducing the standard gage-factor definition:

$$GF = \frac{\frac{\Delta R}{R}}{\varepsilon}$$

and substituting into eq (3):

$$\hat{\varepsilon}_1^1 = (F_a\lambda_{1\,a}\varepsilon_1^1 + F_t\lambda_{1\,t}\varepsilon_2^1)\,/\,GF \qquad (4)$$

where $\hat{\varepsilon}_1^1$ = indicated strain in the 1 direction due to uniaxial stress applied in the 1 direction. In accordance with the normal practice of gage manufacturers, the transverse-gage factor F_t is replaced by $K_t F_a$, where K_t is defined as the 'transverse sensitivity'. Noting also that $\varepsilon_2^1 = -\nu_{12}\varepsilon_1^1$, eq (4) can be rewritten as:

$$\hat{\varepsilon}_1^1 = F_a \varepsilon_1^1 (\lambda_{1a} - \nu_{12} K_t \lambda_{1t})/GF \qquad (5)$$

Similarly, for the gage in the $\underline{2}$ direction,

$$\hat{\varepsilon}_2^1 = F_a \varepsilon_1^1 (-\nu_{12}\lambda_{2a} + K_t \lambda_{2t})/GF \qquad (6)$$

where λ_{2a}, λ_{2t} = axial and transverse strain-transmission coefficients for a gage oriented in the $\underline{2}$ direction. The apparent major elastic modulus is then:

$$E_1^* = \frac{\sigma_1}{\hat{\varepsilon}_1^1} = \frac{\sigma_1 \cdot GF}{F_a \varepsilon_1^1 (\lambda_{1a} - \nu_{12} K_t \lambda_{1t})}$$

But, since $\sigma_1/\varepsilon_1^1 = E_1$,

$$E_1^* = \frac{E_1 \cdot GF}{F_a(\lambda_{1a} - \nu_{12} K_t \lambda_{1t})} \qquad (7)$$

From eqs (5) and (6), the apparent major Poisson's ratio becomes:

$$\nu_{12}^* = \frac{-\hat{\varepsilon}_2^1}{\hat{\varepsilon}_1^1} = \frac{\nu_{12}\lambda_{2a} - K_t \lambda_{2t}}{\lambda_{1a} - \nu_{12} K_t \lambda_{1t}} \qquad (8)$$

The calibration procedure can then be repeated (using the identical gage type) for uniaxial stress applied in the $\underline{2}$ direction as shown in Fig. 3. In this case, the indicated strain for the gage oriented in the $\underline{1}$ direction is:

$$\hat{\varepsilon}_1^2 = F_a \varepsilon_1^2 (-\nu_{21}\lambda_{1a} + K_t \lambda_{1t})/GF \qquad (9)$$

and that for the gage in the $\underline{2}$ direction becomes:

$$\hat{\varepsilon}_2^2 = F_a \varepsilon_2^2 (\lambda_{2a} - \nu_{21} K_t \lambda_{2t})/GF \qquad (10)$$

The apparent minor elastic modulus is then:

$$E_2^* = \frac{E_2 \cdot GF}{F_a(\lambda_{2a} - \nu_{21} K_t \lambda_{2t})} \qquad (11)$$

From eqs (9) and (10), the apparent minor Poisson's ratio is:

$$\nu_{21}^* = \frac{\nu_{21}\lambda_{1a} - K_t \lambda_{1t}}{\lambda_{2a} - \nu_{21} K_t \lambda_{2t}} \qquad (12)$$

When strain measurements are subsequently made on actual test objects in an arbitrary strain field, with

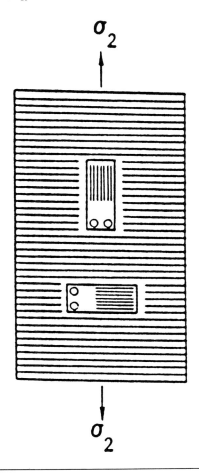

Fig. 3—Transverse specimen for evaluating apparent elastic properties E_2^* and ν_{21}^*

gages of the identical type aligned along the principal material directions, the indicated strains can be expressed as:

$$\hat{\varepsilon}_x = F_a(\lambda_{1a}\varepsilon_x + K_t \lambda_{1t}\varepsilon_y)/GF \qquad (13)$$

$$\hat{\varepsilon}_y = F_a(\lambda_{2a}\varepsilon_y + K_t \lambda_{2t}\varepsilon_x)/GF \qquad (14)$$

Note that subscripts x and y are used in eqs (13) and (14) to designate strains in the $\underline{1}$ and $\underline{2}$ directions, respectively, to avoid confusion with the previously used notation for the calibration strains in the same directions.

Assuming linear-elastic behavior of the test material, and writing the usual orthotropic-normal-stress/normal-strain relationships in terms of the indicated strains and the apparent elastic properties:

$$\hat{\sigma}_x = \frac{E_1^*}{1 - \nu_{12}^* \nu_{21}^*} (\hat{\varepsilon}_x + \nu_{21}^* \hat{\varepsilon}_y) \qquad (15)$$

$$\hat{\sigma}_y = \frac{E_2^*}{1 - \nu_{12}^* \nu_{21}^*} (\hat{\varepsilon}_y + \nu_{12}^* \hat{\varepsilon}_x) \qquad (16)$$

After substituting eqs (7), (8), (11), (12), (13), and (14) into eqs (15) and (16), and reducing,

$$\hat{\sigma}_x = \frac{E_1}{1 - \nu_{12}\nu_{21}} (\varepsilon_x + \nu_{21}\varepsilon_y) = \sigma_x \qquad (17)$$

$$\hat{\sigma}_y = \frac{E_2}{1 - \nu_{12}\nu_{21}} (\varepsilon_y + \nu_{12}\varepsilon_x) = \sigma_y \qquad (18)$$

This result demonstrates that the reinforcement and transverse-sensitivity errors in the indicated strains are cancelled by the corresponding errors in the apparent elastic properties when normal strains are converted to normal stresses using eqs (15) and (16). Although it is common practice in orthotropic mechanics to use the products $\nu_{21}E_1$ and $\nu_{12}E_2$ interchangeably, the same relationship evidently does not hold for the apparent elastic properties. Since the product of eqs (7) and (12) is not equal to that of eqs (8) and (11) in this model of the reinforcement effect, it is necessary that eqs (15) and (16) be applied in the form shown to achieve error cancellation.

The method of compensation for reinforcement and transverse-sensitivity effects proposed here is based on the model generally expressed in eq (2). It implicitly assumes that mechanical interaction effects between gages in the 1 and 2 directions, if present, are the same for the calibration conditions as they are for strain measurement on a test part. To satisfy this condition, a tee rosette (with two grids, 90 deg apart) represents a repeatably convenient means for implementing the method in the compensation of indicated normal strains.

To fully establish the state of stress on the principal material planes, it is also necessary to determine the shear stress, which is related to the shear strain through the shear modulus:

$$\tau_{12} = G_{12}\gamma_{12} \qquad (19)$$

Equation (19) presents a similar opportunity for cancellation of reinforcement and transverse-sensitivity errors by combining indicated strains with an apparent shear modulus.

The American Society for Testing and Materials (ASTM) has established a standard practice for measuring the shear modulus of a unidirectionally reinforced plastic with strain gages.[7,8] The ASTM standard calls for a calibration specimen in the form of a balanced, symmetric, ±45-deg laminate, fabricated from layers of the test material. A tensile specimen is then made from the laminate, and two strain gages are installed, as indicated in Fig. 4.

Fig. 4—Balanced, symmetric, ±45 deg laminate specimen for evaluating apparent shear modulus G_{12}^*

With this construction, the shear stress on the principal material planes is the same for each lamina in the laminate, and is equal to $\sigma_3/2$. Similarly, the shear strain is the same in every lamina.[7] Ignoring reinforcement effects for the moment (as the ASTM standard does), it can be demonstrated that the difference in indicated strains from two gages with their axes 90-deg apart is equal to the shear strain along the bisector of those axes.[8] For the specimen and gage arrangement of Fig. 4, the bisector of the gage axes is a principal material axis, and thus,

$$\gamma_{12} = \varepsilon_3 - \varepsilon_4 \qquad (20)$$

where $\varepsilon_3, \varepsilon_4$ = strains parallel and perpendicular, respectively, to the longitudinal axis of the calibration specimen in Fig. 4.

Applying the previously used reinforcement model

to express the indicated strain in the 3 direction on the calibration specimen,

$$\hat{\varepsilon}_3 = (F_a \lambda_{3_a} \varepsilon_3 + F_t \lambda_{3_t} \varepsilon_4) GF$$

where $\lambda_{3_a}, \lambda_{3_t}$ = axial and transverse strain-transmission coefficients for a gage oriented in the 3 direction. And substituting $F_t = K_t F_a$,

$$\hat{\varepsilon}_3 = F_a(\lambda_{3_a}\varepsilon_3 + K_t \lambda_{3_t} \varepsilon_4)/GF \qquad (21)$$

Similarly, for the indicated strain in the 4 direction,

$$\hat{\varepsilon}_4 = F_a(\lambda_{4_a}\varepsilon_4 + K_t \lambda_{4_t} \varepsilon_3)/GF \qquad (22)$$

From the mechanical symmetry of the gage environments, it can be assumed that $\lambda_{3_a} = \lambda_{4_a}$ and $\lambda_{3_t} = \lambda_{4_t}$. Thus, the indicated shear strain becomes:

$$\hat{\gamma}_{12_c} = \hat{\varepsilon}_3 - \hat{\varepsilon}_4 = F_a(\varepsilon_3 - \varepsilon_4)(\lambda_{3_a} - K_t \lambda_{3_t})/GF \qquad (23)$$

where $\hat{\gamma}_{12_c}$ = indicated shear strain on the principal material axes under calibration conditions. The apparent shear modulus is then:

$$G_{12}^* = \frac{\tau_{12}}{\hat{\gamma}_{12_c}} = \frac{\sigma_3 \bullet GF}{2F_a(\varepsilon_3 - \varepsilon_4)(\lambda_{3a} - K_t \lambda_{3t})}$$

But, $\sigma_3/2(\varepsilon_3 - \varepsilon_4) = G_{12}$, the actual shear modulus of the material. Therefore,

$$G_{12}^* = \frac{G_{12} \bullet GF}{F_a(\lambda_{3a} - K_t \lambda_{3t})} \qquad (24)$$

Subsequently, the same gage arrangement, with the identical gage type, is used to determine the shear strain on an actual test part in an arbitrary stress state. If the strains in the 3 and 4 directions are labeled ε_x and ε_y, respectively, the indicated shear strain on the principal material axes is:

$$\hat{\gamma}_{12} = \hat{\varepsilon}_x - \hat{\varepsilon}_y = F_a(\varepsilon_x - \varepsilon_y)(\lambda_{3a} - K_t \lambda_{3t})/GF \qquad (25)$$

The indicated shear stress is calculated from:

$$\hat{\tau}_{12} = G_{12}^* \hat{\gamma}_{12} \qquad (26)$$

Substituting eqs (24) and (25) into eq (26) demonstrates that:

$$\hat{\tau}_{12} = G_{12}(\varepsilon_x - \varepsilon_y) = G_{12}\gamma_{12} = \tau_{12} \qquad (27)$$

Thus, the errors due to reinforcement and transverse sensitivity are canceled when the shear stress is calculated from the indicated shear strain and the apparent shear modulus as previously measured with the same type of strain gage.

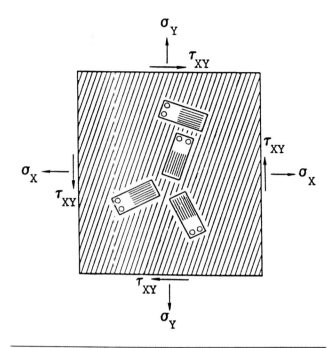

Fig. 5—Strain-gage array for independent measurement of shear and normal strains on the principal material axes of a unidirectionally reinforced composite

Conclusion

A method has been described here for achieving compensation of reinforcement and transverse-sensitivity errors when making strain measurements on an orthotropic material such as a unidirectionally reinforced plastic. The method is applied separately to normal and shear strains to obtain the complete state of stress on the principal material axes. Although not expressly noted in the foregoing, this compensation procedure will also cancel a constant error in gage factor, if present. As a result, the gage-factor control of the instrumentation can be set at any convenient value, as long as it is the same during properties calibration and experimental stress analysis.

In the practical implementation of this method, when compensating for reinforcement effects in both normal and shear strains, four strain-gage grids are required—two along the principal material axes, and the other two at ± 45 deg from one of the axes. Accurate gage alignment is, of course, critical to the procedure. To eliminate possible secondary reinforcement effects of adjacent gages, the gage configuration should be the same in the calibration tests as it is during experimental stress analysis. In other words, if an array of four gage grids is used to determine the complete state of stress for experimental-stress-analysis purposes, the same array should probably be present during all calibrations for elastic properties, whether or not strain measurements are made with the superfluous grids. A further restriction on the physical arrangement of the array, or rosette, is that the two grids used for shear measurement should lie in a mechanically symmetric environment, so that they have the same axial and transverse strain-transmission coefficients ($\lambda_{3_a} = \lambda_{4_a}$ and $\lambda_{3_t} = \lambda_{4_t}$). One such arrangement is indicated schematically in Fig. 5.

It is worth noting that much of the published data on the elastic properties of unidirectionally reinforced plastics was measured with strain gages. Such being the case, these properties may include, in varying degrees, errors due to gage-reinforcement effects if the elastic modulus is low in one or more directions as hypothesized here. When the same properties are employed in the data reduction of strain measurements (also containing reinforcement errors) for stress-analysis purposes, at least partial compensation for the errors must occur by default. Considering the variability in gage stiffness from type to type, however, and pending the quantitative characterization of strain-gage reinforcement effects, the method proposed here seems to offer improved accuracy in the experimental stress analysis of materials conforming to the reinforcement model.

References

1. McCalvey, L.F., "Strain Measurements on Low-Modulus Materials," presented at BSSM Conf., Univ. of Surrey, U.K. (Sept. 1982).
2. White, R.N., "Model Study of the Failure of a Steel Bin Structure," presented at ASCE/SESA Exchange Session on Physical Modeling of Shell and Space Structures, ASCE Annual Conv., New Orleans, LA (Oct. 1982).
3. Stehlin, P., "Strain Distribution In and Around Strain Gauges," J. Strain Anal., 7 (3), 228-235 (1972).
4. Beatty, M.F. and Chewning, S.W., "Numerical Analysis of the Reinforcement Effect of a Strain Gage Applied to a Soft Material," Int. J. Eng. Sci., 17, 907-915 (1979).
5. Measurements Group, Inc., "Errors Due to Transverse Sensitivity in Strain Gages," Tech. Note TN-509 (1982).
6. ASTM, "Standard Practice for Inplane Shear Stress-Strain Response of Unidirectional Reinforced Plastics," Standard No. D 3518-76 (reapproved 1982).
7. Rosen, B.W., "A Simple Procedure for Experimental Determination of the Longitudinal Shear Modulus of Unidirectional Composite" J. Comp. Mat., 6, 552-554 (Oct. 1972).
8. Perry, C.C., "Plane Shear Measurement with Strain Gages," EXPERIMENTAL MECHANICS, 9 (1), 19N-22N (Jan. 1969). (Measurements Group Tech Note TN-512).

Section IIIF

Normal-Stress and Shear-Stress Gages and Rosettes*

by Charles W. Bert

Introduction

Three-element strain rosettes have long been used in experimental-mechanics investigations on isotropic materials. Also in use for some time have been the concepts of 'stress gages' and 'plane-shear gages' in which the gage outputs are proportional to the maximum normal stress and maximum in-plane shear stress, respectively. This paper generalizes all of these concepts for applications to fiber-reinforced composite materials. The data-reduction equations presented explicitly incorporate the effect of strain-gage transverse sensitivity.

The specific objectives of the present section are threefold: (1) determination of strain and stress measurement needs for experiments on composite-material coupons and composite structures; (2) determination of the simplest strain element arrangements to measure the desired quantities; (3) derivation of the data-reduction equations associated with the various arrangements, including explicit incorporation of transverse sensitivity.

Strain-Measurement Needs in Composites

The present investigator envisions three classes of strain-measurement needs associated with composites: (1) material characterization (Ref. 2), in which many times the specimen can be designed so that the principal-strain direction coincides with the fiber direction; (2) transducer applications, in which it may be necessary to determine only a particular stress component, rather than a complete state; (3) composite structures, where the structural complexity (especially lack of symmetry) is such that principal-strain directions are not even known a priori and thus, in general, would not coincide with the fiber direction. Further, in this application, it is necessary to know the complete stress state relative to the material-symmetry axes, since composite-material failure theories depend upon all of these stresses.

Performance of a Single Strain-Gage Element Including Transverse Sensitivity

In the present work, it is assumed that the gage element is sufficiently large relative to the microstruc-

ture of the composite (say the fiber diameter) that the composite may be assumed to be macroscopically homogeneous, although orthotropic with respect to the fiber direction (see Section IIA on 'Anisotropic-Material Behavior'). Further, it is assumed that there are no significant microstructural couple-stress effects, i.e., the stress and strain tensors are assumed to be symmetric.

The composite is assumed to have a known major material-symmetry axis, as is the case for a unidirectionally reinforced layer of material. The composite is assumed to obey Hooke's law generalized to the orthotropic, plane-stress case. This requires that the gage elements be perfectly bonded to the surface of the composite, not embedded in it. It also implies that the composite is perfectly (linearly) elastic. This, of course, implies that the present work is not intended for highly viscoelastic materials, metal-constituent composites loaded into the plastic range, or highly nonlinear ('bimodular') elastomer-matrix composites. Of course, it may be possible to generalize the present work to include some of these complicating effects.

The output of a strain gage in a biaxial-strain field is represented by:

$$\Delta R/R_o = F_a \varepsilon_a + F_t \varepsilon_t \qquad (1)$$

where F_a and F_t are the axial and transverse gage factors, $\Delta R/R_o$ is the unit change in electrical resistance, and ε_a and ε_t are the axial and transverse strains. Here, the terms axial and transverse are relative to the strain element, and as is customary, the gage sensitivity to shear strain is neglected.

The transverse-sensitivity factor is defined as follows:

$$K \equiv F_t/F_a \qquad (2)$$

Eq (1) becomes

$$\Delta R/R_o = F_a(\varepsilon_a + K\varepsilon_t) \qquad (3)$$

This section is primarily based on the author's previous work (Ref. 1), which contains a historical review of the literature on the subject.

In a uniaxial-stress field with the applied stress oriented along the gage element, which in turn is mounted on the major material-symmetry axis (fiber direction) of an orthotropic material (see Fig. 1), one has

$$\varepsilon_t = -\nu_{12}\varepsilon_a \qquad (4)$$

Thus, eq (3) becomes

$$\Delta R/R_o = (1 - K\nu_{12})F_a\varepsilon_a \qquad (5)$$

Since strain-gage manufacturers specify transverse sensitivity when the gage is used on an isotropic material, ν_{12} should be replaced by ν_i, the Poisson's ratio of the manufacturer's calibration material (usually steel, with $\nu_i = 0.285$). Eq (5) may therefore be rewritten as

$$\Delta R/R_o = (1 - K\nu_i)F_a\varepsilon_u \qquad (6)$$

where ε_u is the strain along the gage axis under the manufacturer's calibration conditions.

The manufacturer's uniaxial-gage factor, F_u, is defined as

$$F_u \equiv (\Delta R/R_o)/\varepsilon_u \qquad (7)$$

Substituting eq (6) into eq (7), one obtains

$$F_u \equiv (1 - K\nu_i)F_a \qquad (8)$$

Combining eqs (3), (7) and (8) and solving for ε_u, which is now redesignated as the apparent axial strain, ε_a', one obtains the following expression which was first derived by Baumberger and Hines:[3]

$$\varepsilon_a' = \frac{\varepsilon_a + K\varepsilon_t}{1 - K\nu_i} \qquad (9)$$

It is noted that ε_a' is a function of ε_t as well as ε_a and depends upon the two parameters, K and ν_i, of the strain element.

When a *uniaxial* stress state having its uniaxial stress applied in the fiber direction exists, a single gage element oriented in *the fiber direction* is sufficient to determine the stress in the fiber direction. Then, directions a and b become directions 1 and 2, respectively. Further, due to the uniaxiality of the stress state, $\varepsilon_2 = -\nu_{12}\varepsilon_1$. Substituting into eq (9) and solving for ε_1 yields

$$\varepsilon_1 = [(1 - K\nu_i)/(1 - K\nu_{12})]\varepsilon_1' \qquad (10)$$

Then, due to the uniaxial nature of the stress state, Hooke's law is simply $\sigma_1 = E_1\varepsilon_1$. Thus,

$$\sigma_1 = [E_1(1 - K\nu_i)/(1 - K\nu_{12})]\varepsilon_1' \qquad (11)$$

Before going to the more complicated cases of stress gages, shear gages, and rosettes applied to composites, it is desirable to review the analogous situation for isotropic-material applications. This is done in abbreviated form in Table 1 (see the end of this chapter).

Fig. 1-Single gage element

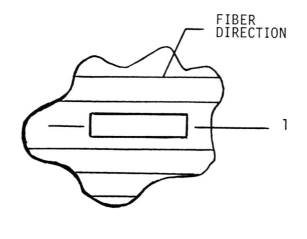

FIBER DIRECTION

1

Performance of an L-Type Rosette

Such a rosette, shown schematically in Fig. 2, has one element oriented in the major material-symmetry (fiber) direction (1) and the other oriented in direction 2 (perpendicular to 1). Applications of eq (9) yield the following expressions for the apparent strains in the 1 and 2 directions:*

$$\varepsilon_1' = \frac{\varepsilon_1 + K_1\varepsilon_2}{1 - K_1\nu_i} \qquad \varepsilon_2' = \frac{\varepsilon_2 + K_2\varepsilon_1}{1 + K_2\nu_i} \qquad (12)$$

Thus, inverting eqs (12), one has

$$\varepsilon_1 = \frac{(1 - K_1\nu_i)\varepsilon_1' - (1 - K_2\nu_i)K_1\varepsilon_2'}{1 - K_1K_2}$$

$$(13)$$

$$\varepsilon_2 = \frac{(1 - K_2\nu_i)\varepsilon_2' - (1 - K_1\nu_i)K_2\varepsilon_1'}{1 - K_1K_2}$$

Since directions 1 and 2 are material-symmetry axes and the stress field is planar, Hooke's law for orthotropic material in plane stress is appropriate (see Section IIA, 'Anisotropic-Material Behavior'):

---

* For a rosette having all gage elements in the same plane, due to different orientations relative to the rolling direction of the foil, K_1 and K_2 may not be the same.

$$\sigma_1 = Q_{11}\varepsilon_1 + Q_{12}\varepsilon_2 \tag{14a}$$

$$\sigma_2 = Q_{12}\varepsilon_1 + Q_{22}\varepsilon_2 \tag{14b}$$

$$\sigma_6 = Q_{66}\varepsilon_6 \tag{14c}$$

where the Q_{ij} are plane-stress reduced stiffnesses, which can be related to the engineering Poisson's ratios and elastic and shear moduli as discussed in Section IIA, 'Anisotropic-Material Behavior'.

Substitution of eqs (13) into eqs (14a) and 14b) yields

$$\sigma_{11} = C_{11}\varepsilon_1' + C_{12}\varepsilon_2'$$

$$\tag{15}$$

$$\sigma_2 = C_{21}\varepsilon_1' + C_{22}\varepsilon_2'$$

and σ_6 cannot be determined in this gage situation. Here,

$$C_{11} = (Q_{11} - K_2 Q_{12})(1 - K_1 v_i)/(1 - K_1 K_2)$$

$$C_{12} = (Q_{12} - K_1 Q_{11})(1 - K_2 v_i)/(1 - K_1 K_2)$$

$$C_{21} = (Q_{12} - K_2 Q_{22})(1 - K_1 v_i)/(1 - K_1 K_2)$$

$$C_{22} = (Q_{22} - K_1 Q_{12})(1 - K_2 v_i)/(1 - K_1 K_2)$$

Fig. 2—L-type rosette

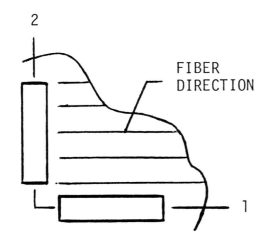

Performance of a V-Type Rosette

This type of rosette, shown schematically in Fig. 3, has two strain elements with a total included angle denoted by 2ϕ. It should be mounted so that one element (denoted by A) is at an orientation $+\phi$ with respect to the major material-symmetry axis (1) and the other (denoted by B) at $-\phi$.

From the strain transformation equations (Mohr's strain circle), one has, for elements A and B:

Fig. 3—V-type rosette

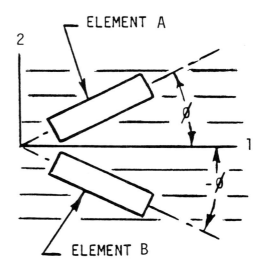

$$\varepsilon_{aA} = \varepsilon_1 \cos^2\phi + \varepsilon_2 \sin^2\phi + \varepsilon_6 \sin\phi \cos\phi$$
$$\varepsilon_{aB} = \varepsilon_1 \cos^2\phi + \varepsilon_2 \sin^2\phi - \varepsilon_6 \sin\phi \cos\phi \tag{16}$$

Adding and subtracting eqs (16), one obtains:

$$\varepsilon_{aA} + \varepsilon_{aB} = 2(\varepsilon_1 \cos^2\phi + \varepsilon_2 \sin^2\phi)$$
$$\varepsilon_{aA} - \varepsilon_{aB} = 2\varepsilon_6 \sin\phi \cos\phi \tag{17}$$

Similarly, one can obtain:

$$\varepsilon_{tA,B} = \varepsilon_1 \sin^2\phi + \varepsilon_2 \cos^2\phi \mp \varepsilon_6 \sin\phi \cos\phi \tag{18}$$

$$\varepsilon_{tA} + \varepsilon_{tB} = 2(\varepsilon_1 \sin^2\phi + \varepsilon_2 \cos^2\phi] \tag{19}$$

$$\varepsilon_{tA} - \varepsilon_{tB} = 2\varepsilon_6 \sin\phi \cos\phi$$

Thus, in view of eqs (9), (17), and (19), the following expressions can be written:*

$$\varepsilon'_{aA} + \varepsilon'_{aB} = \frac{2}{1 - K\nu_i} [\varepsilon_1 \cos^2\phi + \varepsilon_2 \sin^2\phi + K(\varepsilon_1 \sin^2\phi + \varepsilon_2 \cos^2\phi)] \quad (20)$$

$$\varepsilon'_{aA} - \varepsilon'_{aB} = \frac{2(1 + K)}{1 - K\nu_i} \varepsilon_6 \sin\phi \cos\phi \quad (21)$$

It is convenient to rewrite eq (20) as

$$\varepsilon'_{aA} + \varepsilon'_{aB} = \frac{2}{1 - K\nu_1} [(\cos^2\phi + K \sin^2\phi)\varepsilon_1 + (\sin^2\phi + K \cos^2\phi)\varepsilon_2] \quad (22)$$

Comparison of eqs (14a) and (22) shows that

$$\sigma_1 = C_1(\varepsilon'_{aA} + \varepsilon'_{aB}) \quad (23)$$

provided that

$$(\sin^2\phi + K \cos^2\phi)/(\cos^2\phi + K \sin^2\phi) = Q_{12}/Q_{11} \equiv \nu_{21}$$

or**

$$\tan\phi = [(\nu_{21} - K)/(1 - K\nu_{21})]^{1/2} \quad (24)$$

If the transverse sensitivity is neglected, eq (24) reduces to the classical expression. It can be shown that the error in angle due to neglect of K in eq (24) is greater for positive values of transverse sensitivity than negative ones of the same absolute value. Further, the error is greater for smaller values of Poisson's ratio, regardless of the sign of K. In fact, if the transverse sensitivity is positive and greater than ν_{21}, ϕ becomes imaginary. Thus, in such a case, it is physically *impossible* to orient the strain elements so that the gage functions as a stress gage. This unfortunate situation could occur in the case of a gage with high, positive K mounted on an advanced composite material which typically has very low values of ν_{21}. Table 2 lists the error in ϕ as a function of a wide range of values of ν_{21} for two limiting values of K, -0.03 and 0.03.

Typical values of ν_{21} are listed in Table 3. It is emphasized that most composite materials have considerably lower values of ν_{21} (and thus greater error in ϕ) than isotropic materials.

The proportionality constant in eq (23), including the effect of K, is

$$C_1 = (1 - K\nu_i)(Q_{11}/2)(\cos^2\phi + K \sin^2\phi)^{-1} \quad (25)$$

Comparison of eqs (14c) and (21) shows that the shear stress is

$$\sigma_6 = C_6(\varepsilon'_{aA} - \varepsilon'_{aB}) \quad (26)$$

where, in general, the constant of proportionality is given by

$$C_6 = \frac{1 - K\nu_i}{2(1 + K)} \frac{Q_{66}}{\sin\phi \cos\phi} \quad (\phi \neq 0) \quad (27)$$

For ϕ given by eq (24), eq (27) becomes

$$C_6 = \frac{1 - K}{1 + K} \frac{(1 - K\nu_i)(1 + \nu_{21})}{[(\nu_{21} - K)(1 - K\nu_{21})]^{1/2}} Q_{66}/2 \quad (28)$$

For the special case in which the maximum-principal-stress direction coincides with the fiber direction, a single strain element oriented at an angle ϕ given by eq (24) is sufficient. Thus, by definition, the shear strain ε_6 is identically zero; eq (21) shows that ε'_{aA} and ε'_{aB} coincide. Thus,

$$\sigma_1 = C'_1\varepsilon'_{aA} \quad (29)$$

where

$$C'_1 = 2C_1 = (1 - K\nu_i)Q_{11} (\cos^2\phi + K \sin^2\phi)^{-1} \quad (30)$$

Performance of a Three-Element Rectangular Rosette

A rectangular rosette, oriented as shown in Fig. 4, is considered here. Elements A and B are oriented in the orthogonal directions 1 and 2 where 1 is the fiber direction. Element C is oriented at 45 deg to the 1 direction.* Normal strains ε_1 and ε_2 can be calculated easily from the measured apparent strains ε'_1 and ε'_2 using eqs (13). An equation to calculate the engineering shear strain ε_6, however, required further derivation, which follows.

These relations are obtained from eqs (16) with ϕ = 45 deg.

$$\varepsilon_C = (1/2)(\varepsilon_1 + \varepsilon_2 + \varepsilon_6) \quad (31a)$$

$$\varepsilon_{CN} = (1/2)(\varepsilon_1 + \varepsilon_2 - \varepsilon_6) \quad (31b)$$

Here, subscripts C and CN refer respectively to the 45-deg direction and the direction normal to it (-45 deg). The introduction of direction CN is merely a mathematical convenience and a fourth physical gage oriented at $\phi = -45$ deg is *not* required.

Here, it is assumed that the individual strain elements making up the rosette are stacked and that each element has the same transverse sensitivity.
** *With 1 denoting the fiber direction, ν_{12} is the major Poisson's ratio and ν_{21} is the minor one, or $\nu_{21} = (E_2/E_1)\nu_{12}$.*

---

* *In traditional rosette analysis, the 45-deg gage element is usually designated as the second gage.*

Adding eqs (31a) and (31b) and solving for ε_{CN}, one obtains

$$\varepsilon_{CN} = \varepsilon_1 + \varepsilon_2 - \varepsilon_N \qquad (32)$$

Subtracting eq (31b) from eq (31a), one obtains

$$\varepsilon_6 = \varepsilon_C - \varepsilon_{CN} \qquad (33)$$

Now, substitution of eq (32) into eq (33) yields

$$\varepsilon_6 = 2\varepsilon_N - \varepsilon_1 - \varepsilon_2 \qquad (34)$$

Application of eq (9) to gage C yields

$$\varepsilon_C' = \frac{\varepsilon_C + K_C\varepsilon_{CN}}{1 - K_C\nu_i} \qquad (35)$$

Using eq (32) to eliminate ε_{CN}, one can rewrite eq (35) as

$$\varepsilon_C = \frac{(1 - K_C)\varepsilon_C + K_C(\varepsilon_1 + \varepsilon_2)}{1 - K_C\nu_i} \qquad (36)$$

Addition of eqs (13) yields

$$\varepsilon_1 + \varepsilon_2 = \frac{(1-K_1\nu_i)(1-K_2)\varepsilon_1' + (1-K_2\nu_i)(1-K_1)\varepsilon_2'}{1 - K_1K_2} \qquad (37)$$

Substituting eqs (37) into eq (36) and solving for ε_C, one obtains the following expression for ε_C as a function of measured apparent strains ε_C', ε_1', and ε_2':

$$\varepsilon_C = \frac{1}{1-K_C}\left\{(1-K_C\nu_i)\,\varepsilon_C' - \right.$$

$$\frac{K_C}{1-K_1K_2}\left[(1-K_1\nu_i)(1-K_2)\varepsilon_1'\right. \qquad (38)$$

$$\left.\left. + (1-K_2\nu_i)(1-K_1)\varepsilon_2'\right]\right\}$$

$$\varepsilon_6 = \frac{2(1-K_C\nu_i)}{1-K_C}\,\varepsilon_C' - \frac{1+K_C}{1-K_C}\left[(1-K_1\nu_i)(1-K_2)\varepsilon_1'\right.$$

$$\left. + (1-K_2\nu_i)(1-K_1)\varepsilon_2'\right] \qquad (39)$$

Equations (13) and (39) constitute data-reduction expressions for determining the complete in-plane strain state (ε_1, ε_2, and ε_6) as a function of the strain-element readings (ε_1', ε_2', and ε_C').

An early historical work on the use of strain rosettes for composites has been written by Curtis[4]. The recent work of Smith and Hunt[5] is also recommended.

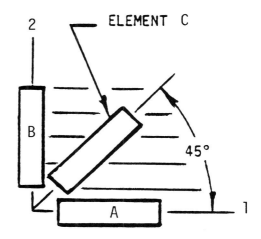

Fig. 4—Three-element rectangular rosette

Performance of a Three-Element Delta Rosette

A delta rosette, oriented as shown in Fig. 5, is now considered. It has been shown by Baumberger and Hines[3] that the actual strains ε_A, ε_B, and ε_C (corrected for rosette transverse sensitivity) are given by

$$\varepsilon_A = \varepsilon_A' - K'(\varepsilon_B' + \varepsilon_C')$$

$$\varepsilon_B = \varepsilon_B' - K'(\varepsilon_A' + \varepsilon_C') \qquad (40)$$

$$\varepsilon_C = \varepsilon_C' - K'(\varepsilon_A' + \varepsilon_B')$$

where ε_A', ε_B', ε_C' are the uncorrected strain readings in strain elements A, B, and C; and K' is a coefficient not to be confused with the manufacturer's transverse sensitivity of the delta rosette; see Ref. 3.

It is assumed here that element A is located in the fiber direction; then $\varepsilon_1 = \varepsilon_A$, and elements B and C are located at $\phi_B = 120$ deg (or -60 deg) and $\phi_C = 60$ deg. Thus, application of eqs (16) gives

$$\varepsilon_B = (1/4)(\varepsilon_1 + 3\varepsilon_2 + \sqrt{3}\,\varepsilon_6) \qquad (41a)$$

$$\varepsilon_C = (1/4)(\varepsilon_1 + 3\varepsilon_2 - \sqrt{3}\,\varepsilon_6) \qquad (41b)$$

Subtracting eq (41b) from eq (41a) yields

$$\varepsilon_6 = (2/\sqrt{3})(\varepsilon_B - \varepsilon_C) \qquad (42)$$

Adding eqs (41a) and (41b) and using $\varepsilon_1 = \varepsilon_A$, one obtains

$$\varepsilon_2 = (1/3)(2\varepsilon_B + 2\varepsilon_C - \varepsilon_A) \qquad (43)$$

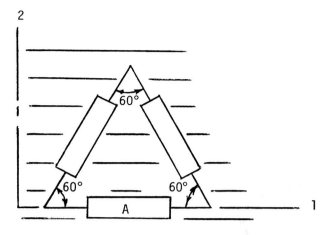

Fig. 5—Three-element delta rosette

2

60°

60° 60°

A

1

Using eqs (40), it can be shown that

$$\varepsilon_1 = \varepsilon_A' - K'(\varepsilon_B' + \varepsilon_C')$$

(44)

$$\varepsilon_2 = (1/3)[\ (2 - K')(\varepsilon_B' + \varepsilon_C') - (1 + 4K')\ \varepsilon_A']$$

$$\varepsilon_6 = (2/\sqrt{3})(1 + K')(\varepsilon_B' - \varepsilon_C')$$

Equation (44) may be used to determine the actual strains in the material-symmetry directons $(\varepsilon_1, \varepsilon_2, \varepsilon_6)$ from knowledge of the delta-rosette readings $(\varepsilon_A', \varepsilon_B', \varepsilon_C')$.

One other type of strain-gage rosette, neither rectangular nor delta in configuration, is the proprietary 'Lambert shear-normal strain gage' (Ref. 6). This rosette contains linear gages perpendicular to each other plus a circular-arc gage.

Special Considerations and Conclusions

Table 4 provides a summary of the various cases for use of strain gages on the surface of composites, while Table 3 lists the corresponding data-reduction equations. It has been shown, for materials meeting the conditions specified, that stress gages and shear gages can be utilized in conjunction with composites. Explicit data-reduction equations have been presented for determining these cases, as well as for the general stress-state case (three-element rosette) of either rectangular or delta configuration.

Potential problem areas in applications of the equations presented are as follows. The minor

Poisson's ration (ν_{21}) is often very small, and thus difficult to measure accurately. Also, the elastic properties may change due to material damage such as transverse cracking. This aspect is beyond the scope of the present investigation.

The V-type stress gage is of limited utility even when used in conjunction with isotropic material. However, for orthotropic material, its use may not be successful for the following reasons: (1) It is not an economic solution, since such a gage, even when available, costs more than two equivalent single-element gages. To obtain such a gage requiring an included angle of, say 20 deg, for use on a composite would be prohibitively expensive. (2) Two separate instrument hookups are required to measure normal stress and shear stress in an arbitrary direction. (3) Errors in the included angle ϕ would be highly amplified. (4) Transverse-sensitivity data for commercial V-type gages are less likely to be available.

Acknowledgements

The author acknowledges the helpful suggestions of Professor Ronald A. Kline of the University of Oklahoma, Dr. C.C. Perry, consulting engineer, Wendell, North Carolina, and Professor Ramesh Talreja of the Technical University of Denmark. The skillful typing by Mrs. Rose Benda is also appreciated.

References

1. Bert, C.W., "Normal and Shear Stress Gages and Rosettes for Orthotropic Composites," EXPERIMENTAL MECHANICS **25**, 288-293 (1985).

2. Bert, C.W., "Experimental Characterization of Composites," Chap. 9, pp. 73-133 in Composite Materials, L.J. Broutman and R. H. Krock, eds., Vol. 8: Structural Design and Analysis, Part II, C.C. Chamis, ed., Academic Press, New York (1975).

3. Baumberger, R. and Hines, F., "Practical Reduction Formulas for Use on Bonded Wire Strain Gages in Two-Dimensional Stress Field," Proceedings, SESA **II**, 113-127 (1944).

4. Curtis, L.F., "Rosettes Show Up the Stresses in Fibre Materials," The Engineer (London) **230** (5949), 33-35 (1970).

5. Smith, H.W. and Hunt, J.P., "Stresses and Strains in Composites from Rosette-Gage Readings," EXPERIMENTAL TECHNIQUES **9** (10), 21-22 (1985).

6. Lambert, W.G., "The LSN Strain Gages," Journal of Engineering Education **59** (1), 74, 76 (1968).

TABLE 1—SUMMARY OF STRAIN-GAGE APPLICATIONS ON ISOTROPIC MATERIALS

Case	Given	Find	Most Economic Solution
1	Uniaxial-stress state with known maximum-principal-stress orientation (x)	σ_{max}	Single gage in x direction
2	Biaxial-stress state with known maximum-principal-stress direction (x)	σ_{max}	Single gage at angle ϕ with respect to x direction
3	Uniaxial- or biaxial-stress state (of unknown principal-stress directions)	σ_x & σ_y	L-type gage
4	Uniaxial- or biaxial-stress state (of unknown principal-stress directions)	σ_x & τ_{xy}	V-type gage
5	Uniaxial- or biaxial-stress state (of unknown principal-stress directions)	σ_{max}, σ_{min}, & τ_{max}	Three-element rosette (rectangular or delta)

TABLE 2—ERROR IN ϕ DUE TO NEGLECT OF K IN EQ. (24), FOR V-TYPE STRESS GAGE

	Error, deg	
ν_{21}	$K = -0.03$	$K = 0.03$
0.03	-3.93	9.83
0.05	-3.18	4.54
0.10	-2.25	2.71
0.15	-1.77	2.10
0.20	-1.46	1.62
0.25	-1.23	1.36
0.30	-1.05	1.15
0.35	-0.99	0.98

TABLE 3—TYPICAL APPROXIMATE VALUES OF THE MAJOR AND MINOR POISSON'S RATIOS FOR VARIOUS MATERIALS

Material	Macroscopic Class	θ_{12}	θ_{21}
Epoxy	Isotropic	0.35	0.35
Aramid/epoxy	Orthotropic	0.34	0.022
Aluminum	Isotropic	0.33	0.33
Boron/epoxy	Orthotropic	0.30	0.030
Steel	Isotropic	0.285	0.285
Graphite/Epoxy (high modulus)	Orthotropic	0.25	0.0063
Graphite/Epoxy (high strength)	Orthotropic	0.27	0.019
Glass/Epoxy	Orthotropic	0.25	0.083
Glass	Isotropic	0.2	0.2
Concrete	Isotropic	0.1	0.1

TABLE 4—SUMMARY OF STRAIN-GAGE APPLICATIONS ON ORTHOTROPIC MATERIALS

Case	Given	Find	Most Economic Solution
1	Uniaxial-stress state with maximum-principal-stress direction known to be along fiber direction (1)	σ_1	Single gage in 1 direction
2	Biaxial-stress state with maximum-principal-stress direction known to be along fiber direction (1)	σ_1	Single gage at angle ϕ with respect to 1 direction
3	Uniaxial- or biaxial-stress state with maximum-principal-stress direction known to be along fiber direction (1)	σ_1 & σ_2	L-type gage lined up with 1 and 2 (perpendicular to 1) directions
4	Uniaxial- or biaxial-stress state (of unknown principal-stress directions)	σ_1 & σ_6	V-type gage with arms at $\pm \phi$ with respect to 1 direction
5	Uniaxial- or biaxial-stress state state (of unknown principal-stress directions)	σ_1, σ_2, & σ_6	Three-element rosette lined up with 1 direction

TABLE 5—LIST OF FORMULAS FOR STRAIN-GAGE APPLICATIONS DISCUSSED IN TABLE 4

Case	Formula	Note
1	Eq (11) in text	Gage parameters k and v_i and composite-material properties E_1 and v_{21} must be known
2	Eq (29) in text	Gage parameters K and v_i and material property Q_{11} must be known; gage element must be oriented at ϕ given by eq (24)
3	Eq (15) in text	Gage parameters K and v_i and material properties Q_{11}, Q_{12}, and Q_{22} must be known
4	Eqs (23) and (26) in text	Gage parameters K and v_i and material properties Q_{11}, Q_{66}, and v_{21} must be known; gage elements must be oriented at $\pm \phi$ given by eq (24)
4	Eqs (14a,b,c,) and and (39) in text for for rectangular rectangular rosette or eqs (14a,b,c,) and (44) for delta rosette	Stress state is arbitrary plane stress; gage element A is oriented in the fiber direction (direction 1)

Section IIIG

Liquid-Metal Strain Gages

by Charles W. Bert

Introduction

Conventional metallic-foil, electric-resistance strain gages are very popular for measuring surface strains and even internal strains (via embedded gages) in composites as well as metals. However, certain composites, notably soft biological tissues and tire cord-rubber, exhibit so little stiffness, due to their highly compliant matrix material, that conventional strain gages cannot be used to take quantitative strain measurements of them. The greater stiffness of the strain-gage material causes a localized stiffening effect which drastically reduces the measured strain magnitudes relative to the actual ones[1]. Furthermore, the strain magnitudes are typically 15 to 20 times greater than those encountered in the more common engineering materials such as metallic alloys. Finally, the poor heat dissipation through cord rubber allows the temperature to build up in the area beneath the current-carrying resistance strain gages. Thus a strain transducer to measure large rubber strains without the difficulties mentioned above is needed.

Clip gages, rubber-wire gages, and liquid-metal strain gages (LMSG or mercury gages) are the three transducers that meet this need.[2,3]. The moire' method with photographic deposition of grids also has been used successfully;[4] however, one encounters difficulties using it on nonplanar surfaces.

The physical principle on which the LMSG is based is the strain-induced change in electrical resistance of a liquid-metal capillary (usually mercury) that is encased within a compliant casing (usually rubber tubing); see Fig. 1. The history of the LMSG concept goes back to the pioneering work of Whitney in 1949, in connection with measuring the circumference of human limbs.[5] In 1953, Whitney[6] reported on further applications in medical science and in the same year Wooley and Hurry[7,8] reported on LMSG use in the rubber industry. The first mechanics-type analysis of the gage to predict its resistance change as a function of strain was provided by Sikorra[9]. His analysis

Fig. 1—Typical dimensions of a liquid-metal strain gage

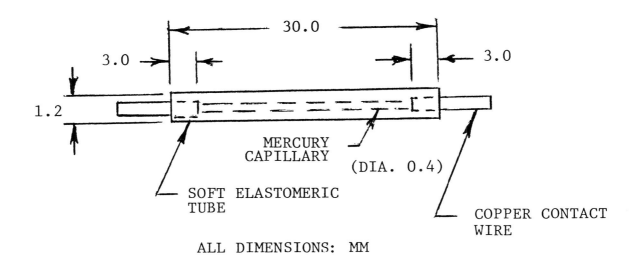

ALL DIMENSIONS: MM

shows that $\Delta R/R_o$ is not linear with strain as it is in the case of ordinary metallic-foil gages. (Further research on the LMSG concept is provided in Refs. 10-14, as well as Refs. 3 and 4.)

The two latest publications relating to LMSG are by Stone et. al.[5] and Bert and Kumar[5]. These describe applications on human knee ligaments and cord-rubber, respectively. The following sections are based primarily upon these two references.

Fabrication and Installation of Gages

The LMSG consists of a column of liquid mercury contained in a compliant tubular casing with lead-wires attached to each end. The steps in fabrication of the gage consist of: (1) either casting of the tubular casing[5] or cutting it from commercially available tubing,[15] (2) injection of liquid mercury into the cavity and insertion of the contact wires at the ends of the cavity, and (3) soldering leadwires to the contact wire. Various materials have been used successfully for the compliant casing, including butyl rubber and silicone rubber. The latter material is advantageous in the filling operation because it is nearly transparent. Capillary diameters ranging from 0.08 mm-0.40 mm have been used.

The most critical aspect of the gage fabrication is filling of the tube with liquid mercury. It is highly recommended that an instrument grade of mercury, low in impurities, be used. Insertion of the mercury is facilitated by use of a syringe; syringe piston should be pushed slowly so that the tube is completely filled by mercury without any trapped air bubbles or pockets. If the compliant tubing is transparent or nearly so, the tubing can be checked for air pockets by visual inspection. A final check of the electrical continuity between the leadwire and the tip of the syringe assures that there are no air bubbles. Next, the syringe tip is removed and that end of the tube is also sealed with a conducting wire of slightly larger diameter than the inside diameter of the tubing.

A wide variety of contact-wire materials, including amalgamated copper and platinum, have been used. Ordinary copper wire is satisfactory; however, there is a corrosion problem, since mercury vigorously attacks copper. Thus, the shelf life an an LMSG may be only a few weeks, although times as long as four months have been reported.[5] When inserting the closing contact wire, one should insert the wire with some axial compression force so as to ensure positive contact with the mercury. It may also be advisable to use a Teflon heat-shrink tube inserted at each end and shrunk by blowing hot air on it, as described in Ref. 15.

After the leadwires are soldered to the contact wires, the gage is ready for a final inspection using an ohmmeter and stretching the gage to approximately 115 percent of its relaxed strength. Continuity should not be broken. The resistance may vary from a fraction of an ohm to several ohms.

To mount the gage, the Teflon end tubes should be bonded to the specimen using a flexible adhesive, such as silicone-rubber adhesive, rather than an epoxy, which has a much higher elastic modulus.

To measure, for example, Poisson's ratio, a transverse gage, as well as a longitudinal one, is needed. For a specimen subjected to uniaxial tension, most materials (those having a positive Poisson's ratio) exhibit compressive strain in the transverse direction. Since a liquid-mercury column contained in a soft capillary tube cannot sustain compressive strain, to measure compressive strain with an LMSG, one must mount the gage with a tensile prestretching that is greater than the largest compressive strain anticipated. Then, as the compression takes place, the prestretching is gradually released. In effect, the use of a longitudinal and a transverse gage constitutes an L-type strain rosette, as described in the chapter on 'Strain-Gage Reinforcement Effects on Orthotropic Materials'. Three-element delta rosettes of LMSG have also been constructed[5]; the data-reduction equations for them are provided in the above-mentioned chapter as well.

Since an LMSG exhibits very low resistance (on the order of an ohm or less), it must be connected to a conventional strain indicator in series with a large resistor (typically 120 ohms).

Calibration and Verification of Gages

The change in resistance of an LMSG is not linear with engineering strain. In fact, as shown theoretically by Sikorra[9] (see also Ref. 5):

$$\Delta R/R_o = F_a (\varepsilon + C\varepsilon^2)$$

where $C \equiv$ conventional linear gage factor, and $\varepsilon \equiv$ engineering strain. Theoretically, $C = 1/2$ and this was borne out by the experiments described by Stone et al.[5] However, Bert and Kumar found the value of C to be much smaller.

Due to the nonlinearity exhibited in eq. (1) with $C \neq 0$), the indicated strain (proportional to $\Delta R/R_o$) is always greater than the actual engineering strain.

An additional test was conducted to verify the validity of the strain measurements (also described in Ref. 15). Strain in a uniaxial specimen (tire cord-rubber with cords at 90 degrees to the loading direction) was measured not only by the LMSG but also by a traveling microscope (Gaertner, having a least count of 0.005 mm) over a 30-mm gage length. The strains measured by the two methods agreed with a mean deviation less than one percent.

Special Considerations and Summary

Liquid-metal strain gages can be fabricated in either single- or delta-rosette configurations. Their main advantages are their low stiffness (essential for use on composites with soft, elastomeric matrices) and high elongation (at least 50 percent). Their principal disadvantages are a short shelf life and a nonlinear calibration curve.

Acknowledgments

The author acknowledges helpful suggestions by Dr. Joseph D. Walter of Firestone Central Research Laboratories and by Professor John L. Turner of the University of South Carolina, as well as the skillful experimental work of his former graduate student, Kumar Shinde, now affiliated with McDonnell Douglas Astronautics Company.

References

1. Beatty, M.F. and Chewning, S.W., "Numerical Analysis of the Reinforcement Effect of a Strain Gage Applied to a Soft Material," Int. J. Eng. Sci., 17, 907-915 (1979).

2. Pugin, V.A., "Electrical Strain Gauges for Measuring Large Deformations," Soviet Rubber Industry, 19 (1), 23-26 (1960).

3. Janssen, M.L. and Walter, J.D., "Rubber Strain Measurements in Bias, Belted Bias and Radial Ply Tires," J. Coated Fibrous Mat., 1, 102-117 (1971).

4. Patel, H.P., Turner, J.L., and Walter, J.D., "Radial Tire Cord-Rubber Composite," Rubber Chem. and Tech., 49, 1095-1110 (1976).

5. Stone, J.E., Madsen, N.H., Milton, J.L., Swinson, W.F., and Turner, J.L., "Developments in the Design and Use of Liquid-Metal Strain Gages," EXPERIMENTAL MECHANICS, 23, 129-139 (1983).

6. Whitney, R.J., "The Measurement of Volume Changes in Human Limbs," J. Physiology, 121, 1-27 (1953).

7. Hurry, J.A. and Wooley, R.P., "A New High-Range Strain Gage," Rubber Age, 73, 799-800 (1953).

8. Wooley, R.P. and Hurry, J.A., U.S. Patent No. 2,739,212 (1956).

9. Sikorra, C.F., "High Elongation Measurements with Foil and Liquid Metal Strain Gages," Inst. Soc. of Amer. Preprint No. 17.11-1-65 (1965).

10. Harting, D., "High Elongation Measurements with Foil and Liquid Metal Strain Gages," Proc., SESA Western Regional Strain Gage Committee, Fall Meeting, 23-28 (1965).

11. Rastrelli, L.V., Anderson, E.L., and Michie, J.D., U.S. Patent No. 3,304,528 (1967).

12. Gregory, R.K., Rastrelli, L.V., and Minor, J.E., "Tire Structural Design Improvement," Aeronautical Systems Div., Wright-Patterson AFB, Ohio, Report ASD-TR-68-12 (1968).

13. Mills, E.J., "A New High-Elongation Strain Gage," presented at the Society for Experimental Stress Analysis Fall Meeting, Indianapolis, Indiana (Oct. 1973).

14. Koogle, T.A., Piziali, R.L., Nagel, D.A., and Perkash, I., "A Motion Transducer for Use in the Intact In-Vitro Human Lumbar Spine," ASME J. Biomech. Eng., 99, 160-195 (1977).

15. Bert, C.W. and Kumar, M., "Measurement of Bimodular Stress-Strain Behavior of Composites Using Liquid-Metal Strain Gages," Experimental Techniques, 6 (6), 16-20 (Dec. 1982).

Section IVA-1

Geometric Moire

by V.J. Parks

Introduction

Displacements and strains can be determined by putting two marks on a surface, measuring the length between them, then loading the body and measuring the length again. The difference between the two lengths is displacement, and the displacement divided by the initial length is the strain. The technique can be expanded by using a series of dots, or intersecting lines, to analyze a whole area. This technique is sometimes called the grid method and is described in detail elsewhere. If the area is large it may be more convenient to take advantage of the fact that such arrays of dots or lines (called gratings), if regular, produce an interference pattern between the loaded and the unloaded array. The pattern is called a moire pattern and is related to the surface displacements in an analyzable way.

The pattern is called moire because of its similarity to a 'watered' silk fabric known as moire. It is called geometric moire here to distinguish it from the similar patterns produced by optical interference of laser beams reflected from very fine line gratings. That method, described elsewhere, is called interferometric moire. Geometric moire can also be used to determine out-of-plane surface displacements, curvatures and slopes. This is done using the shadows and the reflections of the grating. Although presented here for application to composite structures, since the method describes only the geometric behavior of surfaces, it is more general, and the only special details in analyzing composites are in surface preparation, and in noting the constituative relations of the material if stresses are desired.

The most common array used in moire is a grating of equispaced parallel lines, usually in one direction (one-way), but sometimes in two perpendicular directions (crossed). Line widths are usually about equal to the spaces between lines. Line densities vary from about 4 to 400 lines per centimeter (10 to 1000 lines per inch). The less dense gratings are available (both one-way and crossed) from commercial art dealers. Denser gratings used by photoengravers, for halftone work, can be obtained from their suppliers. The densest gratings are sold by optical equipment suppliers. All these gratings are on transparent material, usually photographic film or glass.

In-Plane Displacements and Strains (In-plane Moire)

To measure in-plane displacements and strains with moire, the surface of the composite should be reasonably smooth and flat. The grating should be one of the denser gratings mentioned above. Usually a grating on film, emulsion side out, is cemented to the surface. Occasionally, the grating is printed on the surface by a "lithographic" process, and indeed the surface is sometimes actually etched, to photoengrave the grating on the surface. Other approaches such as drawing, scribing, or stamping are possible.

Unless the material is translucent, there must be visible contrast between the grating lines and the intermediate spaces. The contrast can be provided by a light surface (e.g., a white silicone rubber cement), flat shiny spaces between rough grooves, white chalk-filled grooves between dark photoengraved lines, etc. If, as in most cases, displacements and strain are needed in more than one direction, a cross grating may be applied to the surface to obtain two sets of orthogonal displacements. An alternative to a cross grating is a set of two identical specimens with one-way gratings in perpendicular directions. Two specimens are sufficient to obtain displacements and strains for all in-plane directions.

Once the specimen grating (SG) is applied, a second grating often of the same density, called the reference grating (RG), is placed over the specimen grating. If the surface is horizontal, the RG may simply be laid on the SG. The RG, usually a photocopy, is either on glass or film, and is placed emulsion side towards the SG, so as to put the two emulsions in intimate contact. Because the surface may not be exactly flat, and will also undulate slightly under load, an RG on film is often better than an RG on glass. In any case a thin film of oil between the SG and RG improves the contrast of the moire pattern. If the surface is not horizontal, the oil film, used to improve contrast, may also be used to support the RG. Otherwise, some sort of fixture may be needed to keep the RG in intimate contact with the SG during loading. The RG should be initially set up with the RG grating lines parallel to and between the SG lines, so as to generate a dark field with no pattern or a minimum pattern.

In principle, the RG should not move once it is set. In practice, the RG is often allowed to move, or is even removed, and is reset at some convenient position after loading. For strain analysis the movement of the RG corresponds to the negative rigid-body movement of the specimen, which does not alter strains. Note that motion of the RG (specifically rotation) will give different patterns that do alter the intermediate analysis.

If the SG is a crossed grating, the RG may be either crossed or one-way. The one-way RG has the advantage of greater contrast and clarity in the pattern. A possible disadvantage is that the two RG settings must be set so as to be exactly 90 degrees with respect to each other. The advantage of the crossed RG grating is that it gives both patterns in one loading. Figure 1 shows sketches of some typical moire setups.

As the body is loaded, a moire pattern is seen to be generated between the two gratings. This pattern is usually recorded photographically. Often photographs are taken at several levels of load, including the unloaded pattern, to provide extra data and note any residual fringes or nonlinear behavior. If the RG is one-way, the specimen must be unloaded, the RG rotated 90 degrees, and the body reloaded and rephotographed to obtain the complimentary set of moire patterns. This completes an experimental sequence and provides all the data necessary for analysis.

Analysis

The first step of the analysis is ordering (or counting) the fringes in the pattern. If the RG is held stationary while loading, it may be possible to watch, and count from zero, those fringes passing through a point during loading. If there is a point on the surface that does not move, the stationary fringe at that point can be taken as a zero fringe. Since displacements are continuous, the fringes are continuous, and can be counted from the known fringe. Unless the actual direction of motion can be observed, this counting does not establish the fringe sign. To do this it is still necessary to determine in which direction the displacements and fringes increase, or decrease. Sometimes this is obvious from the overall motion of the surface. An indirect method to establish the fringe direction is to shift the RG perpendicular to its grating lines. If the fringes move in the direction of the shift, it indicates a positive direct derivative and the displacements and fringes must increase in the positive direction of the corresponding coordinate. This is true regardless of which direction is chosen as the positive coordinate direction. And vice-versa, if the fringes move opposite to the shift, the fringes must increase in the direction opposite to the arbitrarily chosen coordinate direction. Figures 1 and 2 illustrate ordering of fringes.

If the RG and SG are both of the same frequency before loading, then the displacements are,

$$u = p\,n_x$$
$$v = p\,n_y$$

where n_x and n_y are the moire fringe orders associated with fringe patterns of the gratings perpendicular to the x and y directions, respectively, p is the reference grating pitch (the reciprocal of the grating frequency) and u and v are the x and y components of the specimen displacement at the point at which n_x and n_y were obtained. Since displacements are vectors, the total displacement at any point is obtained from

$$U = \sqrt{u^2 + v^2}$$

and the angle of the direction of motion is

$$\Theta = \arctan u/v$$

If strains are desired, they can be obtained by differentiating the Cartesian displacements in the Cartesian directions, as shown in Fig. 2. This gives four derivatives at each point on the surface analyzed:

$$du/dx \qquad du/dy \qquad dv/dx \qquad dv/dy$$

The strains on the surface are geometrically related to these four derivatives.

$$\varepsilon_{xx} = \sqrt{1 + 2(du/dx) + (du/dx)^2 + (dv/dx)^2} - 1$$

$$\varepsilon_{yy} = \sqrt{1 + 2(dv/dy) + (du/dy)^2 + (dv/dy)^2} - 1$$

$$d_{xy} = \arcsin \left\{ [du/dy + dv/dx + (du/dx)(du/dy) + (dv/dx)(dv/dy)] / (1 + \varepsilon_{xx})(1 + \varepsilon_{yy}) \right\}$$

where ε_{xx} and ε_{yy} are the strains in the x and y directions and d_{xy} is the shear strain. The derivatives du/dy and dv/dx are called cross derivatives and are equal to the tangents of the angles of rotation of the x and y directions, respectively. If both these derivatives are small with respect to the direct derivatives, du/dx and dv/dy, the above equations reduce to

$$\varepsilon_{xx} = du/dx$$

$$\varepsilon_{yy} = dv/dy$$

$$d_{xy} = du/dy + dv/dx$$

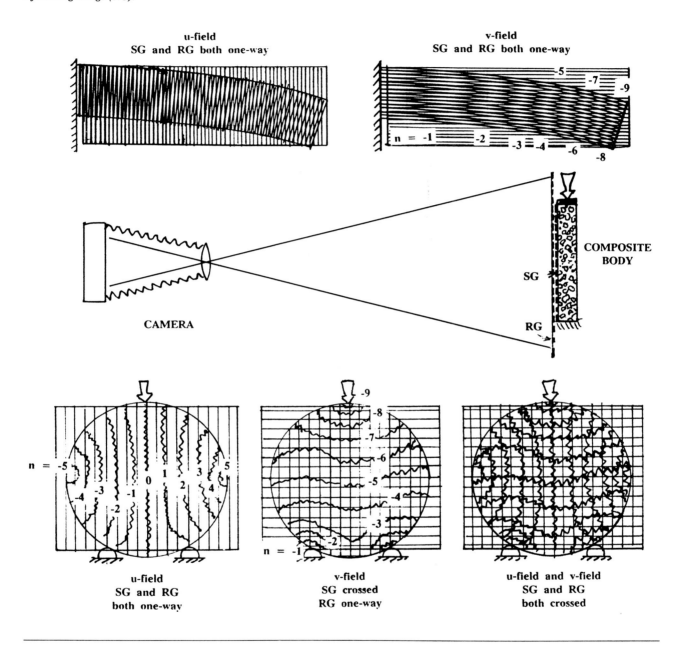

Figure 1—Basic in-plane moire camera set-up, and various combinations of cross and one-way specimen gratings (SG) and reference gratings (RG).

If both the x and y directions have zero rotation (as on an axis of symmetry), the reduced equations are valid regardless of the strain magnitude. On the other hand, if the x and y directions are, for example, the axial and circumferential directions of a torsion bar, regardless of how small the strains are, the more general equations are needed because the tangent of the angle of rotation in the axial direction (the cross derivative dv/dx) is larger than the strain. Usually, the choice of which equations are to be used can be decided by comparing results of both sets for several data points. If the results are different, the more general equations are correct.

If the strains are small or the region of interest is small, there may be too few fringes to obtain accurate derivatives. The interferometric moire method described elsewhere will provide many more fringes. Alternatively there are two approaches to increase fringes with the geometric moire, by multiplication or by addition. To multiply fringes, an RG with a frequency that is an integer multiple of the unloaded SG frequency is used (twice the SG, or three times the

59

59

SG, etc.). Two or three times the fringes will be obtained. The RG pitch, which is one-half or one-third the SG pitch, must then be used in the above equations for the analysis.

To add fringes, the RG will have a frequency a few percent different from that of the SG. Such gratings are usually produced by photographing the grating a few percent larger and smaller than the original. If a 1000-line-per-inch grating is enlarged or reduced one percent, when mounted on the unstrained original, 10 fringes per inch are obtained. If an SG, equal in frequency to a 1000-line-per-inch RG, is subjected to strain and produces two fringes per inch, then for the same load, with the one-percent grating difference between the RG and the unstrained SG, either 8 or 12 fringes will be produced depending on whether the

---

Figure 2—Fringe-vs-position plots used to obtain the four spacial derivatives of displacements.

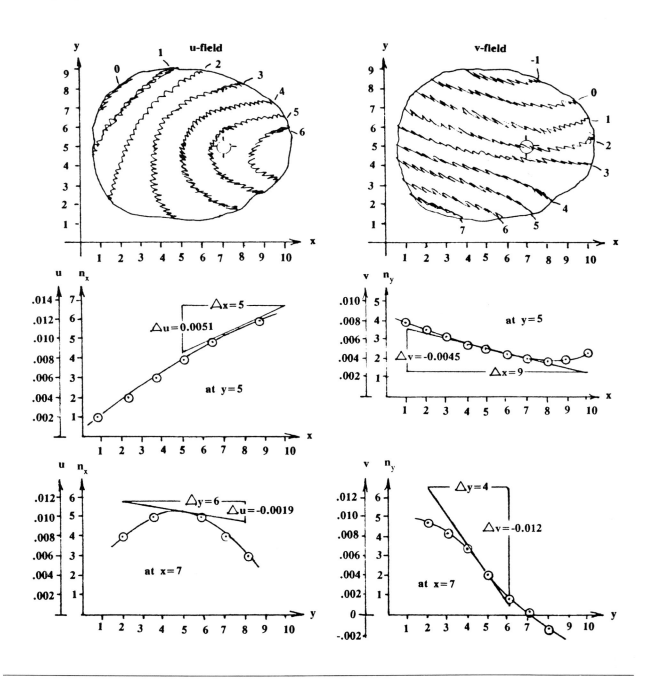

60

actual strain was positive or negative. This addition technique sometimes is called mismatch, because of the mismatched RG.

Example

Figure 2 shows portions of typical u-field and v-field moire-fringe patterns in a region of a specimen to be analyzed for strain along the horizontal line shown. The x and y coordinates are shown in the conventional manner. The direction of increasing fringe orders must be established by: (1) watching as the patterns develop, (2) knowledge of the overall motion, or (3) by shifting the RG over the loaded specimen and noting the direction of fringe motion.

If strains only are required, any fringe can be assigned an integral order and all other fringes given consecutive values in the direction found, as suggested above. If actual displacements are required, the absolute value of some fringe in the pattern must be determined knowing a specific displacement (often zero at a fixed point or on a line of symmetry) or by watching the fringe grow from zero load. The sort of fringe orders obtained following these procedures are shown in Fig. 2.

Fringe vs. position can then be plotted either by measuring the actual position of integral fringes on the line and plotting, or by estimating the fractional fringe order at equal positional intervals along the line and plotting. In Fig. 2, the u fringes are plotted with integral fringes, and the v fringes at unit intervals, as examples of the two methods of plotting. A displacement scale is added to each graph by simply multiplying the fringe coordinate scale by the pitch p. The displacement scale should be given the same units as the position scale. The slope of the curves at any position (the unitless quotient of the change in displacement divided by the change of position) gives the displacement derivatives at that position.

The four plots in Fig. 2 give a complete set of four derivatives to obtain the complete two-dimensional strain tensor at the point ($x = 7, y = 5$). If strains are required along the whole x line, then a series of short plots, across the x line in the y direction can be made, such as the two cross plots of u vs. y and v vs. y shown in Fig. 2. These, along with the other two plots, give a complete set of four derivatives all along the x line. In the case of a full-field analysis, the fringes are plotted all the way across the specimen in both directions. In such full-field analyses, it is usually more convenient to record fractional fringes at equal intervals, so the same readings can be used for both the x and y plots.

If there is an initial mismatch (to generate more fringes by addition), the direct derivative of the mismatch pattern on the unloaded body (typically a single value over the whole field) is subtracted from the direct derivatives of the patterns on the loaded body point by point to determine the true derivatives. The cross derivatives do not need adjustment since the initial mismatch pattern does not have cross derivatives.

Principal strains and their directions follow directly from the Cartesian components using the usual transformation equations. Either Cartesian or principal stresses and their directions may be determined from the constitutive equations of the composite material being analyzed.

Slope and Curvature (Reflection Moire)

Testing

In order to measure slope and curvature with moire, the surface of the composite must be, or must be made, sufficiently reflective, to image one of the lower density gratings described above. If the surface is not reflective it may be polished, or coated with a reflective coating, or even temporarily wet with a reflective liquid. A grating is then placed at some distance d in front or to the side of the surface as shown in Fig. 3. The reflected image in the composite surface is photographed. This first photograph serves as a reference grating (RG) image. The composite is loaded and the grating image rephotographed on the same film to obtain a specimen grating (SG) image. The right-hand arrangement in Fig. 3 shows how an RG and an SG can be obtained in a single exposure with the use of an additional grating. This arrangement is also convenient if the topography of an unloaded surface is desired. On loading the body, the reflected image of the grating lines moves due to the change in slope (rotation) of the composite surface, in planes perpendicular to the grating lines. Thus the double exposure produces a pattern of fringes proportional to the change in surface slope, in the direction perpendicular to the grating lines. If the surface was initially flat, then the fringes represent the final surface slope. By rotating the grating 90 deg. in its plane and repeating the procedure, another double exposure photograph is obtained, representing a second set of slopes, perpendicular in the surface plane to the first set. The two sets of slope are sufficient to describe all slopes on the surface (just as two in-plane moire patterns give the complete in-plane displacement field), and allow determination of all curvatures.

Analysis

The slope and curvature analyses follow closely that of the displacement and strain analyses, respectively, of in-plane moire. The orders of the fringes are first determined. If the pattern is obtained from a double exposure, order cannot be obtained by watching the pattern develop as with in-plane moire. However, the slopes are usually visible, and often there is a peak or valley on the surface containing a zero fringe, from which counting can begin. The sign of the slope is arbitrary, and often unimportant, as with in-plane displacements. To obtain the derivative of slopes (curvature), however, positive x and y directions are arbitrarily assigned, and using a zero fringe, sequential orders are assigned to consecutive fringes. The direction of increasing orders will determine the sign of curvature, and can be reversed to change the sign of curvature. The sign of curvature is conventional; for example, a concave curvature is often given a negative sign and a convex curvature a positive sign.

The angles of rotation are obtained from the fringe orders. The angle of rotation is proportional to the fringe order, to the grating pitch and to the reciprocal of twice the distance between the grating and the surface being analyzed (as shown in Fig. 3). The slope (or change in slope if the surface is not initially flat) is then

$$\theta_x = p\, n_x/2d$$

$$\theta_y = p\, n_y/2d$$

where n_x and n_y are fringe orders produced by gratings perpendicular to the x and y directions, respectively, p is the grating pitch, and θ_x and θ_y are the respective slopes expressed in radians. The vector sum of these two slopes gives the maximum slope at

Figure 3—Two arrangements of components that will generate the moire patterns of surface slope.

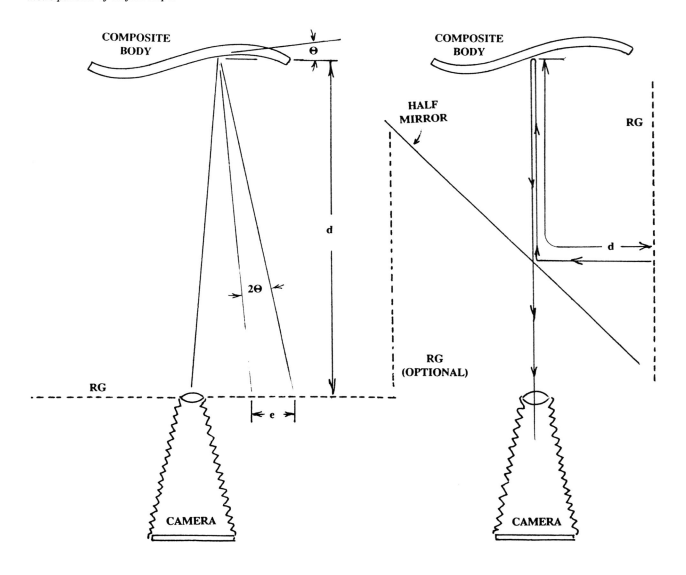

Figure 4—Fringe-vs-position plots illustrating method to obtain curvature of a surface. The fringes correspond to the slopes in the two Cartesian directions, and the derivatives of the fringes correspond to, the two Cartesian curvatures, and the two warps (derivatives of slopes in the direction perpendicular to the slope direction).

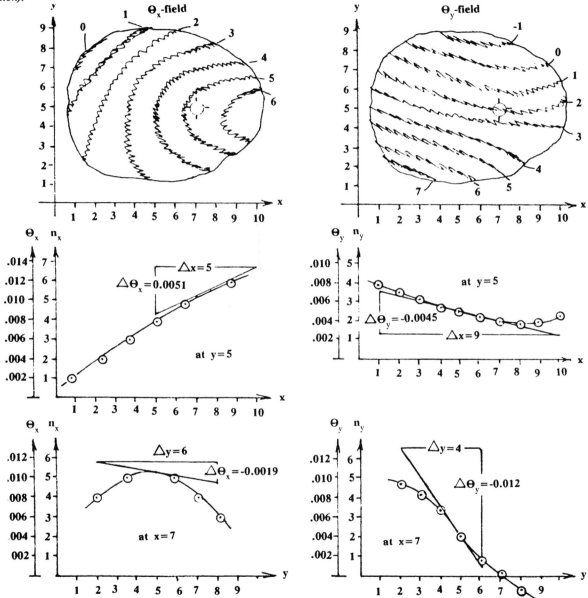

that point on the surface. Curvature is the derivative of slope. Using K to represent curvature,

$$K_{xx} = d\theta_x/dx$$
$$K_{yy} = d\theta_y/dy$$

And, similar to in-plane strain analysis, cross derivatives (sometimes called warp) can be written

$$K_{xy} = d\theta_x/dy$$
$$K_{yx} = d\theta_y/dx$$

These four curvature derivatives can be obtained in the same manner as the in-plane derivatives of displacement. They can be obtained at a point, along a line or for the whole field of the reflection-moiré patterns.

Example

Figure 4 repeats the illustrations and procedures shown in Fig. 2, but applied to slopes and curvatures.

63

Topography and Out-Of-Plane Displacements (Shadow Moire)

Testing

To measure the out-of-plane displacements or topography of the surface of a composite with moire, a grating's shadow is cast on the surface. The composite surface is often painted with white tempera, or a flat white paint, so that the shadow has optimum contrast. The densities of the gratings used in shadow moire are the lowest of those mentioned in the introduction. The grating is placed directly in front of and parallel to the surface, even touching the surface at its highest points. The light used to produce the shadow should be sufficiently far from the grating and surface to produce sharp shadows of the lines and is usually a point or line source. A line light source must be set parallel to the grating lines. By photographing the grating and its shadow on the surface, a fringe pattern is obtained that is approximately proportional to the distance of the surface from the grating. A flat grating produces a fringe pattern of topographic, or constant-height, contours. If the surface was flat before loading, these contours correspond to the out-of-plane displacements of the surface due to load. This same fringe pattern can be obtained by projecting a grating from a collimnated light source, and viewing through a similar grating.

Analysis

If the viewing direction or the camera axis is approximately perpendicular to the grating plane and the composite surface, then the distance along the grating surface between the point at which a line's shadow is viewed and a point on the line itself is related to the distance between the grating surface and the composite's surface. If the angle at the composite surface between the camera axis and the axis of the light beam is termed a, then, as shown in Fig. 5, $\tan a = e/h$, where h is the distance from the grating plane to the composite surface, and e is the distance between a grating line and its shadow.

If the surface is touching the grating, any shadow will coincide with the line (or lines) that produced it and there will be no fringe. However, if the gap h between the grating and the composite surface is such

---

Figure 5—Two arrangements of components to obtain surface topography, or out-of-plane displacements of a loaded plate.

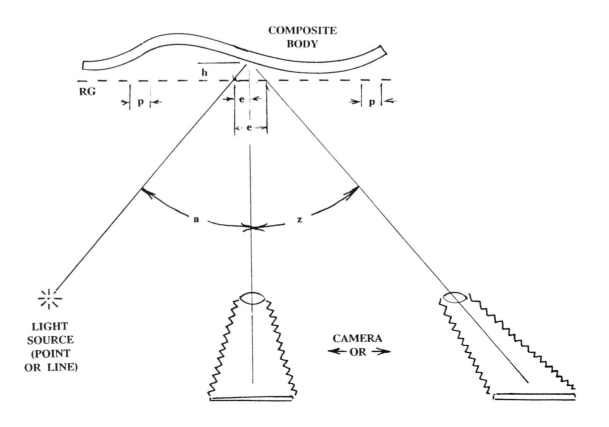

as to shift the shadow half a pitch from the line (or lines) that produced it, the lines and their shadow will generate a dark fringe as viewed by the camera. This fringe area is associated with a half-pitch shift and can be designated as a half-fringe order. Thus the distance e introduced above is seen to be equal to np where p is the grating pitch and n represents shadow-moire fringe orders, with zero and subsequent integers specifying light moire-fringe orders. Dark shadow-moire fringes are then ordered as 1/2, 1-1/2, 2-1/2, and so on.

The equation representing the out-of-plane displacement or topography of the surface can be written

$$h = n \, p \, / \tan a$$

If both camera and light source are at angles a and z respectively from the grating normal, similar reasoning gives a similar expression.

$$h = n \, p \, / \, (\tan a + \tan z)$$

Arrangements for using both these expressions are shown in Fig. 5.

Unlike the in-plane displacement moire, and the slope patterns of reflection moire, the shadow moire requires just one pattern (not two) to describe the out-of-plane displacements. However, it can be dif-
ferentiated like the others. The spacial derivative of out-of-plane displacement is slope, and differentiating in two Cartesian directions gives a complete expression of slope (the same results as obtained directly from reflection moire). In principle, the parallel could be carried further by differentiating the slopes, obtained from the shadow moire, to obtain the four curvature derivatives. However, the errors inherent in double differentiation seriously limit this approach.

Example

Figure 6 is the shadow-moire pattern of a mannequin bust made with a 50-line-per-inch grating and a fluorescent light bulb masked with a 1/8 in. slit at 80 in. from the grating and 20 deg. off the camera axis. The formula gives a contour constant of

$$h = n \, (.02) \, / \tan 20 \text{ deg.}$$

$$h = 0.055 \, n \text{ in.}$$

The 22 fringes between the tip of the chin and the corner of the mouth indicate a depth of over one inch.

References

General

Dally, J.W. and Riley, W.F., *Experimental Stress Analysis*, McGraw-Hill, Ch. 12. (1978).

Durelli, A.J., and Parks, V.J., *Moire Analysis of Strain*, Prentice-Hall (1970).

Parks, V.J., "Strain Measurement Using Grids," *Optical Eng.*, 21(4), 633-639 (1982).

Kobayashi, A.S. Ed., *Handbook on Experimental Mechanics*, Prentice-Hall, Ch. 6 and 7 (1987).

Kobayashi, A.S. Ed., *Manual of Engineering Stress Analysis*, Prentice-Hall, Ch. 6 (1982).

Sciammarella, C.A., "The Moire Method—A Review." EXPERIMENTAL MECHANICS, 22 (11), 418-433 (1982), and Discussion, 23 (12), 446-449 (1983).

In-plane Moire

Chiang, F-P., Parks, V.J. and Durelli, A.J., Moire-Fringe Interpolation and Multiplication by Fringe Shifting, EXPERIMENTAL MECHANICS, 8 (12), 554-560 (1968).

Post, D., "Moire Grid-Analyzer for Stress Analysis." EXPERIMENTAL MECHANICS. 5 (11), 366-377 (1965) and Discussion, 6 (5), 287-288 (1966).

Post, D., "Sharpening and Multiplication of Moire Fringes," EXPERIMENTAL MECHANICS, 7 (5), 154-159 (1967).

Reflection Moire

Kao, T.Y. and Chiang, F-P, "Family of Grating Techniques of Slope and Curvature Measurements for Static and Dynamic Flexure of Plates," *Optical Eng.*, 21 (4), 721-742 (1982).

Ligtenberg, F.K., "The Moire Method: A New Experimental Method for the Determination of Moments in Small Slab Models," SESA, 12 (2), 83-98 (1954).

Shadow Moire

Halioua, M., Krishnamurthy, R.S., Liu, H. and Chiang, F-P, "Projection Moire with Moving Gratings for Automated 3-D Topography," *Applied Optics*, 22 (6) 850-855 (1983).

Pirodda, L. "Shadow and Projection Moire Techniques for Absolute or Relative Mapping of Surface Shapes," *Optical Eng.* 21 (4), 640-649 (1982).

Takasaki, H. "Moire Topology," *Appl. Optics.* 9 (6) 1457-1472 (1970) and 12 (4), 845-850 (1973).

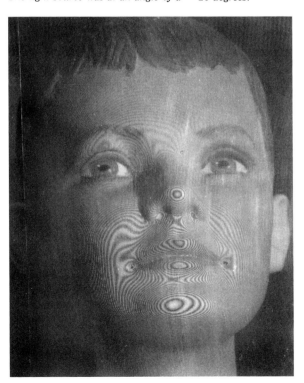

Figure 6—Shadow moire on a mannequin obtained with a 2 line/mm grating, and the set-up (center camera) shown in Fig. 5. The light source was at an angle of a = 20 degrees.

Section IV A-2

Moire Interferometry for Composites

by Daniel Post

Introduction

Moire interferometry is used to measure in-plane displacements, U and V, on flat surfaces. It extends the sensitivity of moire methods into the subwavelength range, making it suitable for the analysis of localized deformations of structural composites. The method is characterized by a remarkable set of qualities: it is a whole-field and real-time method, and it provides high sensitivity, excellent fringe contrast, high spatial resolution and extensive range. Moire interferometry nicely fills the gap between capabilities of other experimental techniques, inasmuch as it can be used in zones of very high strain gradients and it can be used to determine shear strains as readily as normal strains.

Fig. 1—(a) N_y or V field in a multi-ply graphite-epoxy tensile specimen with a central hole; hole diameter is 6.3 mm (0.25 in.).

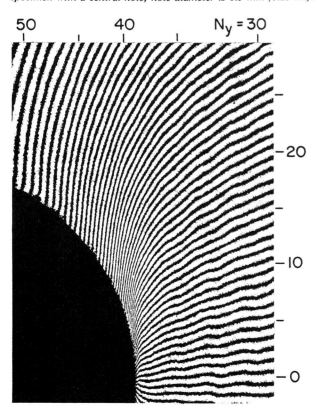

(b) Influence of load level on normal (ε) and shear (γ) strain concentration factors; P_f is failure load

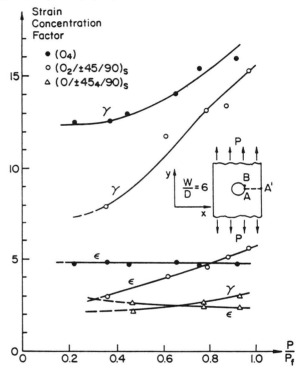

For composites, moire interferometry can be used for measurements in both macromechanics and micromechanics domains. Figure 1 shows an example of macromechanics from a study of strain concentrations around a hole in different laminates[1]. The extremely high spatial resolution capability is illustrated by the concentration of fringes near point B, the point of maximum shear strain. The graph shows how the normal and shear strain-concentration factors vary with load level for three different laminates.

Figure 2 illustrates a micromechanics problem. The boron fibers in this metal matrix specimen were 0.14 mm in diameter. Plastic slip zones occurred in the aluminum matrix on both sides of the marked fiber[2]. The zig-zag appearance of fringes are real deformation effects, not optical noise of the experiments. They signify smaller displacement gradients in the fibers, and larger gradients in the more compliant aluminum matrix. Currently, micromechanics capabilities extend to the fiber level of coarse-fiber composites and to the ply level of fine-fiber composites.

This chapter is intended as a tutorial for the laboratory practice of moire interferometry. Instructions and general guidelines are included for introductory practice. Theoretical aspects, advanced aspects and some details are not included. For these, the investigator should refer periodically to the *Handbook on Experimental Mechanics*[3] and the technical literature as his/her experience and understanding matures.

Fig. 2—N_y or V field in a boron-aluminum tensile specimen with a central slot. Stacking sequence [0/±45]_s. Slot length 5.5 mm (0.22 in.)

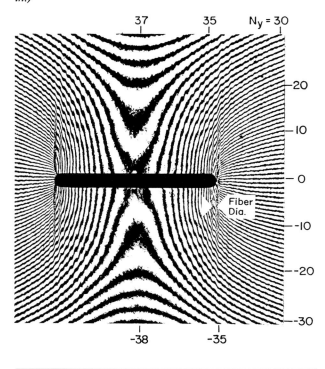

Brief Description of the Method

Suitable optical arrangements are illustrated schematically in Fig. 3. The specimen is prepared in advance with a diffraction grating of frequency $f/2$ lines/mm (l/mm). A parallel beam of laser light arrives at the specimen at angle $-a$, while a portion of the same beam is redirected by a plane mirror to arrive at the symmetrical angle $+a$. Since they are mutually coherent, the two incident beams generate an interference pattern comprised of alternating bright bands (constructive interference) and dark bands (destructive). The bands are very closely

Fig. 3—Two-beam optical system in which the collimating element is (a) a parabolic mirror and

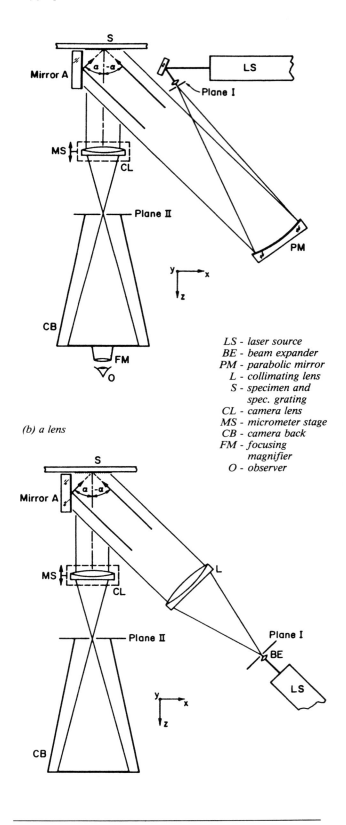

(b) a lens

LS - laser source
BE - beam expander
PM - parabolic mirror
L - collimating lens
S - specimen and spec. grating
CL - camera lens
MS - micrometer stage
CB - camera back
FM - focusing magnifier
O - observer

68

spaced and function as a reference grating. It is not a real or physical grating, but instead, a virtual reference grating. Its frequency is f lines (bright bands) per mm, which is related to wavelength λ and angle a by

$$\sin a = \frac{\lambda f}{2} \qquad (1)$$

Typically, $f = 2400$ l/mm $= 60{,}960$ l/in., as in the case of Figs. 1 and 2. The corresponding initial or no-load frequency of the specimen grating is 1200 l/mm.

The specimen grating and virtual reference grating interact to form a moire pattern. This is photographed by the camera, which is focused on the plane of the specimen. The moire is (essentially) a null field (uniform intensity across the field, devoid of fringes) before loads are applied to the specimen. After loading, the moire-fringe pattern seen in the camera is a contour map in which the fringe order at any point is proportional to the in-plane displacement at the point. Their relationship is

$$U = \frac{N_x}{f} \quad \text{or} \quad V = \frac{N_y}{f} \qquad (2)$$

In Fig. 3, the grating lines are perpendicular to the x axis; the moire senses U, i.e., the x component of displacement; the fringe orders are denoted by N_x. If the specimen (and coordinate axes) is rotated by 90 deg, then the N_y system of fringes (the V field) is recorded.

The normal strains ε_x and ε_y on the surface of the specimen, and the shear strains γ_{xy}, can be determined from the U and V displacement fields by the small strain relationships

$$\varepsilon_x = \frac{\partial U}{\partial x} = \frac{1}{f}\left(\frac{\partial N_x}{\partial x}\right)$$

$$\qquad (3)$$

$$\varepsilon_y = \frac{\partial V}{\partial y} = \frac{1}{f}\left(\frac{\partial N_y}{\partial y}\right)$$

and

$$\gamma_{xy} = \frac{\partial U}{\partial y} + \frac{\partial V}{\partial x} = \frac{1}{f}\left(\frac{\partial N_x}{\partial y} + \frac{\partial N_y}{\partial x}\right) \qquad (4)$$

For strain analysis, the important parameters are the rates of change of fringe orders in the x and y directions—or the x and y components of the fringe gradient—or the local density or spacing of the fringes.

Specimen Grating

Typically, crossed-line gratings are used, having 1200 l/mm perpendicular to both the x and y axes. They are phase-type gratings, so the common terminology, 'lines', really means grooves or furrows. For grooves of sinusoidal cross section, the depth of the grooves and the reflectivity of the grating surface determines the diffraction efficiency of the grating. Maximum efficiency corresponds to maximum light energy in the moire pattern seen in the camera.

Specimen gratings are usually produced by a replication process. Basically, this consists of casting a thin film of a plastic adhesive (e.g., an epoxy or acrylic) between the specimen and a special mold. The mold has the high-frequency grooves of the diffraction grating. When the mold is removed, the specimen is left with a thin film of adhesive firmly bonded to it, with the grooves of the specimen grating in its outer surface.

Molds

The molds used for the moire patterns shown here were made by a holographic process on photographic plates (Kodak High Resolution Plate type TE). The process is described in the *Handbook*[2]. The depth of the groove cannot be controlled and optimized with this process and only about 10 percent of the maximum efficiency is achieved. Nevertheless, excellent moire patterns are produced. The quality of the moire pattern depends more on the uniformity of the grooves than on their depth. The implication of reduced efficiency is merely the need for increased exposure time or increased laser power for recording the moire pattern.*

Grating Alignment

Before replication, some means must be provided to align the grating lines with respect to the specimen. This is usually done by means of an alignment bar as illustrated in Fig. 4(a). The bar A is first aligned with respect to the mold B and cemented to it. Then the bar rests along the edge of the specimen for grating-to-specimen alignment.

A convenient alignment fixture is illustrated in Fig. 4(b). It consists of a flat base C, a bar D with a narrow groove parallel to the base and an adjustable fixture E. The mold is attached temporarily to the fix-

---

* *Work on producing high-efficiency molds is underway in the author's laboratory, using photoresist as the recording medium instead of photographic plates. We will assist investigators with current information and possibly provide molds for their use. Please contact the author.*

ture with double-sided adhesive tape. An unexpanded low-power laser beam intercepts the grating mold as shown. The fixture E is adjusted so that the +1 and −1 order diffracted beams strike the parallel groove. This makes the grating lines accurately parallel to the base C. Next, a thin parallel bar A (e.g., 1/16 in. thick by 1/2 in. wide) is cemented to the mold, while the bar rests on the base as a reference surface; Fig. 4(c) illustrates this step. A quick-setting cyanoacrylate cement is effective. Care must be taken to avoid forming a filet of cement along the alignment edge of the bar, since that would affect the alignment accuracy.

Replication

Surface preparation for composite specimens depends upon the nature of the surface. If it is reasonably smooth, cleaning with solvents is sufficient. If the surface is deeply textured, for example by a scrim cloth impression, some smoothing is needed. For smoothing, hand grinding with fine abrasive paper is satisfactory. An alternative is to fill the valleys in the surface with a plastic. A good procedure is to pour a small puddle of room-temperature-curing liquid epoxy on the specimen, cover it with an acrylic plate, and pry the acrylic off after the epoxy has cured; no release agent is needed, since acrylics do not bond well to most epoxies.

The replication process is illustrated in Fig. 5 for molds made on photographic plates. First, the mold is soaked in a dilute solution (1:100) of Kodak Photoflo for one minute, rinsed and dried. This aids the mold separation later in the replication process. Then a highly reflective metallic film of aluminum is applied by evaporation (also called vacuum deposition). The mold should not be heated in advance of the aluminum deposition, as that increases adhesion of aluminum to the mold. Following the steps of Fig. 5, a small puddle of liquid adhesive is applied to the specimen surface. The aluminized mold is brought into contact with the adhesive and pressed to squeeze the adhesive into a thin film. The mold is moved to correctly position the alignment bar with respect to the specimen. The mold can slide easily along the specimen, and its position is fixed by adhesive tape or by weights placed along its edges. A weight is placed on the mold to continue to squeeze out the excess adhesive. The weight should apply one to five pounds per square inch of specimen-contact area to achieve a grating thickness of about 0.001 in.

Various adhesives can be used. A room-temperature epoxy that performs well is PC10-C.* When the epoxy gels (before it becomes hard), the weight should be removed and the excess epoxy should be separated from the specimen by cutting

---

* Sources of components are listed under References. Similar components are available through other sources. The list does not infer preference, superiority or economy of these products relative to others.

Fig. 4—Specimen grating alignment. (a) Use of the alignment bar. (b) Alignment fixture. (c) Alignment bar cemented to mold.

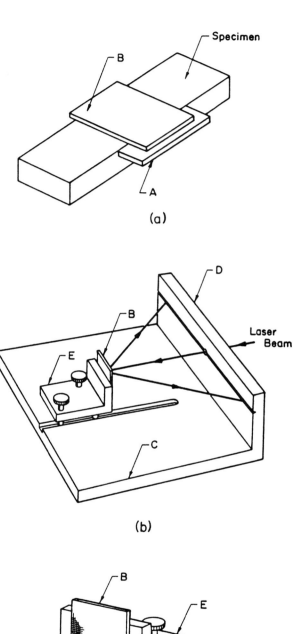

(a)

(b)

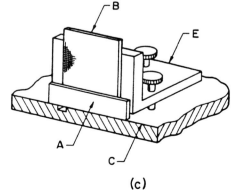

(c)

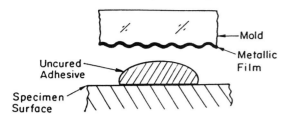

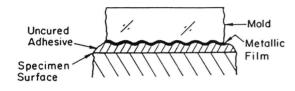

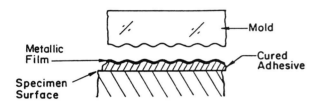

Fig. 5—Steps in producing the specimen grating by a casting or replication process; the reflective metallic film is transferred to the specimen grating. Best results are obtained when the metallic film is freshly applied by vacuum deposition, e.g., within 48 hours.

along the specimen edges with a sharp razor knife. Then the epoxy is allowed to harden fully before the mold is pried off. The weakest interface is between the mold and the aluminum film, so the reflective aluminum is transferred to the specimen grating in the replication process. The specimen grating is ready for use without additional trimming.

A note of caution regarding environmental conditions: multidirectional laminated composites exhibit thermal stresses and strains when temperature changes are encountered. To avoid this effect, the room temperature at which the moire patterns are recorded should be the same as the temperature that prevailed when the grating was replicated. With composites that absorb moisture from the atmosphere, the specimen should be kept at its moisture-equilibrium humidity for the time between the replication of the specimen grating and the recording of the moire patterns.

Curved Surfaces

The methods described here cannot be applied to the general analysis of deformations of curved sur-

faces. Local deformations can be determined, however, insofar as the local portion of the surface can be approximated by a flat surface. For a convex surface the specimen grating can be applied in the normal way, but with the flat mold positioned tangent to the local region of interest. Of course, the thickness of the specimen grating would increase with distance from the tangent point and the analysis should be restricted to the zone where the thickness is sufficiently small. For concave surfaces, the flat mold must be cut to a narrow or small size to prevent excess thickness of the specimen grating.

While the above procedure is adequate for many problems, there is scope for other schemes. Examples include the replication of curved or flexible grating molds, and the direct generation of gratings on the specimen.

The Moire Interferometer

Size of Field

For composites, the region of interest is always (or nearly always!) rather small. This is because the ply thickness cannot be scaled up readily, so we do not use an enlarged model of our specimen. In addition, the strain gradients are often so severe in composites that we are interested in detailed observation of a local zone. Figure 1 is a good example, where the hole radius is 1/8 in. and the region of interest is smaller than that.

A field size of one inch or smaller is recommended. The small size makes the apparatus reasonably compact and efficient in utilization of light energy. It reduces the investment in apparatus and the problem of environmental control, and leads to quicker success. The excellent fringe contrast and spatial resolution inherent in these techniques permits great enlargement of the moire-fringe patterns.

Basic Two-Beam System—Components

Figure 3 illustrates two versions of the optical system. The only difference is the element that produces the collimated (or parallel) beam of laser light. A 5-mW helium-neon laser is suitable for initial work. Higher power is convenient when large magnification is desired in the camera screen and when a four-beam system is used. A polarized laser should be used, with its plane of polarization perpendicular to the plane of the diagram.

The next element is a beam expander. This is merely a short focal-length positive lens; a microscope lens is usually used. Microscope lens powers from 5x to 20x are suitable. The angle of the cone of light emerging from the lens increases with increased power. Too large an angle would be wasteful since it would spread the light beyond the field of the collimating lens or mirror. A pinhole is sometimes used at the node, or apex of the cone of light from the lens. The combination of a pinhole and microscope lens is called a spatial filter. Its purpose is to filter out noise (or nonuniform intensity) in the beam. The source of noise is primarily dust on the microscope lens which causes diffraction and interference rings in the beam. Such noise is not highly detrimental in moire interferometry, so use of the pinhole is not recommended for initial work. Instead, the microscope lens should be maintained as clean as possible.

The collimating element can be either a lens or a parabolic mirror. In either case, the speed of the element (its focal length divided by its diameter) should be five or larger. High optical quality is desired for this element to control the wavefront distortion of the collimated beam. It is clear, referring to Fig. 3, that the collimated beam must be wider than the field size. If $f = 2400$ l/mm and $\lambda = 633$ mm, angle a is 49.4 deg; then, the collimated beam should be at least 1.5 times the field size, or 1.5 in. in diameter for a 1-in. diameter field. Parabolic mirrors accurate to $\lambda/8$ are readily available at low cost. When a parabolic mirror is used, the angle between the central ray that strikes the mirror and the reflection of this ray should be minimized. A high-quality collimating lens of 1.5-in. aperture is relatively expensive, but it becomes more competitive for smaller field sizes.

The plane mirror A that directs half the beam onto the specimen grating should be flat to $\lambda/8$. It should be mounted on a fixture that allows fine adjustment of its angle about horizontal and vertical axes. (The parabolic mirror should be mounted on such a fixture, too.)

The camera lens should have an aperture that is larger than the field size, so that it can accept off-axis rays from the specimen. A photographic camera lens, an enlarger lens, or an achromatic doublet are all good choices if the large aperture condition is satisfied. Magnification of the image of two to four times is suggested. The lens parameters can be calculated by the simple lens formulas

$$M = \frac{d_i}{d_o} \qquad (5)$$

$$\frac{1}{FL} = \frac{1}{d_o} + \frac{1}{d_i} \qquad (6)$$

where M is magnification, d_o is the distance between the specimen grating (object) and the lens, d_i is the distance between the camera-film plane (image) and the lens, and FL is the focal length of the lens. It is useful to mount the lens on a micrometer stage to facilitate accurate focusing of the specimen surface onto the film plane.

The author finds a 4 x 5 sheet-film camera most convenient. It can be a standard 4 x 5 camera, used without the lens. Alternatively, it can be the camera back, alone, which accepts the sheet-film holder, plus a cardboard enclosure. Kodak Technical Pan film and Kodak Contrast Process Pan film are excellent choices because of their high resolution and good contrast. Faster films can be selected if shorter exposure time is needed.

A focusing magnifier of 5x to 10x power is recommended for critical focusing. The magnifier should be adjusted to the eye of the investigator. This is done by making fine marks on the *inside* surface of the camera screen and adjusting the magnifier for sharpest focus of the marks.

Loading System

Loading fixtures that are specially designed or adapted for each specimen are recommended. This permits mounting the fixture on the same table as the optical apparatus, thus minimizing vibration problems. In general, the fixture should be massive and stiff. Fixed displacement screw-loading fixtures are best, with loads measured by a strain-gage transducer. Dead-weight loading devices have proved troublesome because of swinging of the weight.

When using a two-beam optical system, one must rotate the loading fixture through 90 deg to observe both the U and V fields. This is done by connecting the loading fixture to a massive supporting plate by means of a single shaft (e.g., a shoulder bolt) with its axis passing through the center of the field. Micrometer-controlled stops should be arranged to facilitate rotational adjustments and repeatable 90-deg rotations.

Optical Table

Vibration control is important, even though it is not as critical as with other interferometric systems. Anti-vibration provisions are most important for the specimen and the plane mirror A. Relative motions of a fraction of the virtual grating pitch ($1/f$) cause loss of fringe contrast. These elements should be linked together through rigid connections. The next most critical are the illuminating optics, since the moire fringes are highly sensitive to the angle of incidence a. Again these should be linked together, but the tolerance here is several wavelengths of relative motion. Least critical is the camera, which can tolerate numerous wavelengths of mechanical motion.

The most practical arrangement is to firmly mount all these components on a stiff, massive table top. The table top can be supported by vibration-damping material such as plastic foam or rubber, or by commercial vibration dampers of various degrees of sophistication. Where the budget allows, a commercial optical table with pneumatic vibration control is convenient.

Air currents within the optical path between the beam expander and specimen can cause the moire fringes to dance. If control is needed, air currents can be attenuated by baffles that protect this optical path, or by a box surrounding the entire system. Rigid plastic foam is a good material for baffles.

Alignment Procedure

The alignment procedure involves the following steps.

(1) Prior to critical alignment of the optical system, set up the components as illustrated in Fig. 3, with angle a approximately 49 deg. This is the angle required for $\lambda = 633$ nm (helium-neon laser) and f = 2400 l/mm; for other conditions, determine a by eq (1). Position a white card with a small hole ($\approx$ 0.1-in. diameter) in plane I and temporarily locate a white card in plane II.

(2) Temporarily install a plane mirror in the beam between the collimator and the specimen and adjust its angle to reflect a bright dot back onto plane I. Adjust the distance between the collimator and the card to minimize the size of the spot. This is called autocollimation and it assures a sufficiently parallel beam. Remove the temporary mirror.

(3) Observe plane I while adjusting the angle of mirror A. Two bright dots will appear in plane I. Adjust the mirror until the two dots are superimposed upon each other and are located near the aperture. This brings mirror A perpendicular to the plane of the specimen grating.

(4) Observe two bright dots on plane II. They will have the appearance illustrated in Fig. 6. Axes x' and y' are parallel to the x and y axes defined in Fig. 3. Rotate the specimen in its plane, together with the loading fixture, until the dots lie on the x' axis. This adjustment moves the dots vertically in opposite directions and makes the lines of the specimen grating parallel to those of the virtual reference grating.

(5) Adjust angle a until the two dots merge and become one. Small adjustments of a can be made by translating the parabolic mirror perpendicular to its axis, or by rotating it about a central axis parallel to the y axis. If a collimating lens is used, a can be adjusted by translating the lens perpendicular to its optical axis.

(6) Remove the card from plane II. Adjust the lateral positions of the camera lens and the camera back for concentricity with the beam coming from the specimen. Adjust them longitudinally for the desired magnification and critical focus of the specimen plane on the film plane.

Special hints are useful here. Two fine marks should be scribed on the specimen grating (use a sharp razor knife) before it is installed in the loading fixture. The marks can be small crosses that will appear in the camera image. These serve two functions: as targets for critical focusing and as gage marks to accurately establish the magnification factor. A focusing magnifier (previously adjusted for critical focus on the inside of the ground-glass camera screen) should be used to assess the focus of the marks on the screen. As an additional hint, the moire fringes should not be used to critically adjust the focus. The fringes appear crisply defined even when the specimen is not properly focused, but the fringe pattern changes with change of focus, especially in regions of high-strain gradient.

Moire fringes should appear on the screen. If they do not, the dots in plane II should be checked for superposition. Fine adjustments can be made with mirror A to obtain a null field. Rotation of mirror A about an axis parallel to y regulates the fringe gradient in the x direction (the fringes of extension). Rotation about an axis parallel to z regulates the fringe gradient in the y direction (the fringes of rotation). These adjustments will cause the two dots in plane I to separate and move relative to the aperture. This is acceptable if the angular separation of the dots is not large, e.g., within 0.05 radians as measured from the collimating element; if it is too large, the dots should be brought together again by adjusting mirror A and repeat the previous step for superposition of two dots in plane II.

The initial or no-load pattern will be a true null field (devoid of fringes) only if the specimen grating is perfect and the virtual reference grating is formed by perfect optics. In general, it will exhibit a small number of fringes across the field. These will be negligible compared to the load-induced fringes in most cases. Otherwise, the fringe order in the initial field should be subtracted from that in the final field for every point of interest.

If desired, a carrier pattern of rotation can be added to the no-load fringes by rigid-body rotation of the specimen and loading fixture. A carrier pattern of extension can be added by changing angle a.

Limitations of Two-Beam Systems

The need for accurate 90-deg rotation of the specimen and loading fixture carries certain penalties. One is that the loading fixture must be

Fig. 6—Two bright dots on plane II are brought together by interferometer adjustments

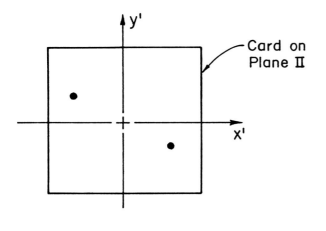

capable of rotation and this limits the simplicity of its design. Another relates to determination of strains. It will be shown in the section *Interpretation of Fringe Patterns* that small errors of rotation have negligible influence on the determination of normal strain, but introduce large errors for shear strains. One means for circumventing the problem with a two-beam system is to use an auxiliary specimen grating, described in Ref. 3, page 346. Another means, one that eliminates the problem, is to use a four-beam optical system.

Four-Beam Optical System

Figure 7 illustrates a four-beam optical system in which the two-beam arrangement of Fig. 3 is duplicated in the horizontal plane and the vertical plane. A beam splitter, *BS*, directs a portion of the collimated beam downwards to mirror *M3*, which in turn directs it to the specimen and to mirror *M4* at an angle *a* in the vertical plane. If an opaque card is placed in the beam on the reflection side of the beam splitter, the transmitted beam strikes the specimen and mirror *M2* and generates the N_x pattern (the *U* displacement field). If the card is placed on the transmission side of the beam splitter, the reflected beam is effective and it generates the N_y pattern (or the *V* field). For each field, the alignment procedure is the same as that for a two-beam system.

Polarization of light incident on the specimen grating should be either parallel or perpendicular to the grating lines. The multiple reflections for the *V* field introduce some ellipticity into the polarization, which diminishes the contrast of N_y moire fringes. To cope with this, a polarizing filter, *P*, is introduced in the focal plane of the camera lens, with its polarizing

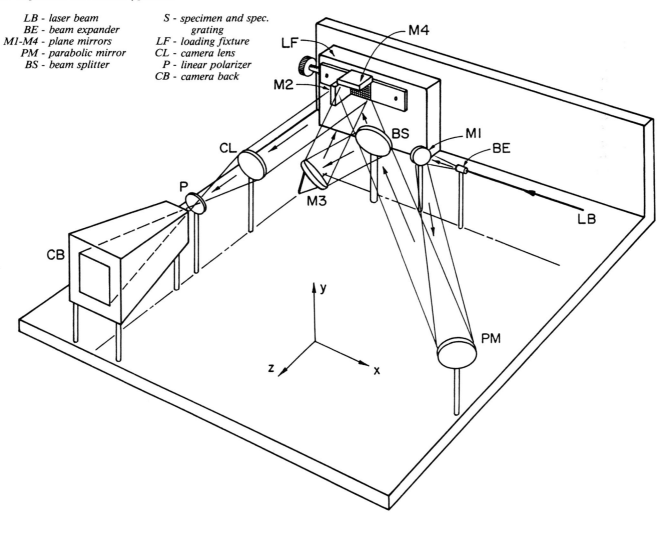

Fig. 7—Four-beam moire interferometer that utilizes a beam splitter to produce the N_x and N_y patterns

LB - laser beam
BE - beam expander
M1-M4 - plane mirrors
PM - parabolic mirror
BS - beam splitter
S - specimen and spec. grating
LF - loading fixture
CL - camera lens
P - linear polarizer
CB - camera back

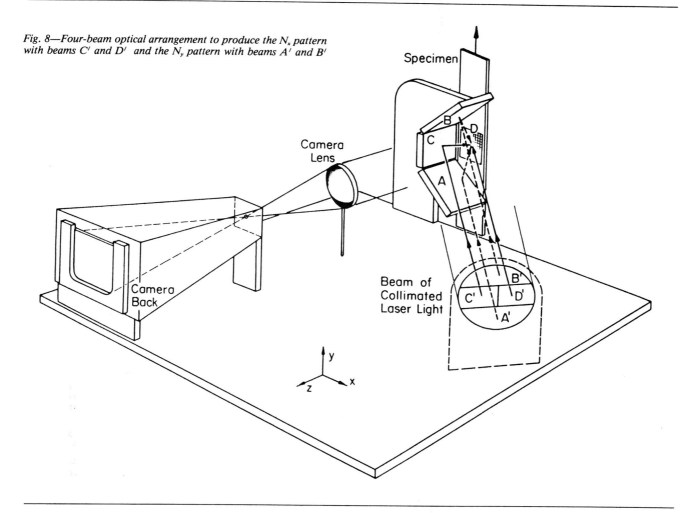

Fig. 8—Four-beam optical arrangement to produce the N_x pattern with beams C' and D' and the N_y pattern with beams A' and B'

axis parallel to the y axis. A polarizing filter laminated in glass, of the sort used on camera lenses, is suitable.

Another four-beam system is illustrated in Fig. 8'. The two 45-deg mirrors are adjusted to direct portions of the collimated beam at angles $+a$ and $-a$ in the vertical plane, to form the N_y or V field. The V field appears in the camera screen when portions C' and D' of the incident collimated beam are blocked; the U field appears when portions A' and B' are blocked. Adjustment for the V field is basically the same as that already described. As before, two dots appear in the focal plane of the collimator (plane I in Fig. 3) and they must be superimposed and coincident (approximately) with the aperture; also, two dots must be superimposed in the focal plane of the camera lens (plane II).

The three-mirror system of Fig. 8' has the advantage that the optical paths are less complicated than those of Fig. 7, and no polarizing filter is required. It has the disadvantage that a larger collimated input beam is required, which means that the system will be less compact and a larger optical table will be needed. Cost of optical elements is not an issue, since the larger collimator will not cost more than a smaller collimator plus a beam splitter. A minor disadvantage of the compact system (Fig. 7) is that the beam splitter might cause optical noise in the form of ghost fringes depicting its optical thickness variation. A proper antireflection coating on the back side of the beam splitter would help, but a better solution is to order a beam splitter that has a strong wedge angle between its front and back faces, such that light from the unwanted reflection passes out of the optical system.

Loading the Specimen

While the procedure for the moire test is simple, various related comments will be offered here. The investigator will (1) install the specimen in the loading fixture; (2) adjust the moire interferometer for null fields (U and V) and photograph the null fields; (3) determine the signs of fringe gradients in the null fields; (4) apply loads to the specimen and photograph the U and V fields; (5) determine the signs of fringe gradients; (6) unload the specimen.

With a 2-beam interferometer, step 2 requires adjusting stops on the rotatable loading fixture to enable its return to the same x and y positions. In the event that the number of fringes in the (so-called) null fields is not negligible, the fields should be photographed and the signs of fringe gradients should be determined. The signs must be determined, too, for the full-load fringe patterns (step 5).

Sign of Fringe Gradient

Knowledge of the signs of the fringe gradients is necessary to correctly establish the fringe orders in the moire-fringe pattern. To determine the sign at any point in the N_x pattern, one should first establish the coordinate system and the positive direction of x, then while observing the fringe pattern in the camera screen, apply a tiny rigid-body displacement to the specimen, for example by pressing lightly against the loading frame. If the displacement is in the positive x direction, the fringe orders will increase at every point in the field. Accordingly, the fringes will move in the direction of the lower fringe orders; the direction of positive fringe gradient (i.e., the direction of increasing fringe orders) is opposite to the direction of fringe movement. The scheme applies to the whole field and it applies, similarly, to the N_y field.

Out-of-Plane Displacements

When a specimen is loaded, it experiences out-of-plane displacements together with in-plane displacements. Fortunately, moire interferometry is virtually insensitive to the out-of-plane motions. The N_x field is completely insensitive to the translations of the specimen in the z direction, and it is completely insensitive to specimen rotations about any axis parallel to x. For a rotation ψ about any axis parallel to y, an extraneous fringe gradient F_e is added to the load-induced gradients, where

$$ F_e = - \frac{f\psi^2}{2} \qquad (7) $$

Angle ψ is the x component of the out-of-plane slope of any point in the specimen, relative to its slope in the initial (no-load) pattern. The change of slope caused by deformation of the specimen would generally be very small and F_e would be negligible. However, it is possible that an inadvertent rotation of the loading fixture occurs when the load is applied, or that the specimen rotates relative to the fixture.

In such a case, the potential error can be corrected. Referring again to Fig. 3, an angular displacement ψ of the specimen (i.e., rotation about a y axis) will cause a separation of two bright dots in plane I. One of two plans can be followed. In plan one, the loading fixture is moved to bring the two dots back to their no-load positions. In plan two, the angle ψ can be measured to calculate F_e and a corresponding cor-

rection can be applied when analyzing the load-induced fringe pattern. Angle ψ is equal to half the change of separation of the two dots divided by the focal length of the collimating lens or mirror. The extraneous fringe gradient is always negative, corresponding to an extraneous compressive strain in the specimen.

In-Plane Rigid-Body Rotations

Rigid-body rotations of the specimen (relative to the virtual reference grating) introduces fringes that are perpendicular to the grating lines. The rotation changes the fringe order of every point in the N_x field by a linear function of y. The gradient $\partial N_x / \partial x$ is not changed.[*] Similarly, $\partial N_y / \partial y$ is unaffected by rotations. Thus, accidental rigid-body rotations can be tolerated, without consequence, for determination of normal strains ε_x and ε_y.

A rigid body rotation of the specimen generates strong cross-derivatives that are mutually dependent; their relationship is

$$ \frac{\partial N_x}{\partial y} = - \frac{\partial N_y}{\partial x} $$

When these contributions of pure rotation are introduced into the shear-strain equation, eq (4), their effect is nil.

The four-beam interferometer capitalizes on this relationship. With it, the x and y virtual reference gratings are fixed in space and the magnitude of the rigid-body specimen rotation, relative to each of them, is identical. Accidental rigid-body rotations of the specimen can result from artifacts of the loading system, but when a four-beam interferometer is used, the rotation has no influence on the calculated shear strains.

With a two-beam interferometer, it is necessary to rotate the specimen and loading fixture through 90 deg to obtain the N_x and N_y fields. Unless the rotation is precisely controlled, it will introduce an error in one cross derivative and an error in the calculated shear strain. The method of an auxiliary reference grating can circumvent the problem.

Trouble Shooting

In the event the moire pattern has poor contrast, one of three causes is likely: polarization, coherence or vibrations. The laser light must be polarized either parallel or perpendicular to the grating lines when the optical scheme of Fig. 3 is used. Otherwise, there will be a loss of fringe contrast, or even total washout of the fringe pattern. As a test, one should check whether the contrast improves when a polarizing filter is inserted in the camera (see Fig. 7); if contrast

---

* There is a second-order influence that becomes meaningful for large rotations. See Ref. 5.

improves, and if the laser can be rotated about its optical axis, one should rotate it to adjust the plane of polarization.

If one zone of the moire pattern has good contrast, but the contrast diminishes with distance from that zone, coherence length could be the problem. The optical system should be arranged to minimize the difference of path lengths of the two rays that strike a point on the specimen grating. In Fig. 3, for example, the two rays that strike the specimen at a point near mirror A have nearly equal path lengths. The two rays that strike a specimen point most distant from mirror A have the greatest difference between path lengths. The path difference must be a small fraction of the coherence length of the laser to retain good contrast. Information on coherence length is available from the laser vendor. In the case of certain high-power lasers, an internal etalon is needed to produce sufficient coherence length.

Vibrations sometimes cause enough fringe motion to mask the visibility of the moire fringes. Mirror A and the specimen are the most critical elements. Vibrations can usually be damped out by pressing a plastic foam against the fixture. Sometimes it is necessary to alter the stiffness or mass of the mountings to change their natural frequencies.

Recording the Moire Pattern

Video vs. Photographic Film

Photographic films have already been recommended for recording the moire pattern. The merits of a video-recording system should be addressed, too. Advantages of video include its comfortable viewing and multiperson viewing. With sophisticated systems, it offers certain opportunities to filter out noise, to digitize the intensity and position data in the moire pattern, to provide hard copies with minimal effort, and to mesh with an automatic or semi-automatic fringe-analysis system. Disadvantages relate primarily to the limited resolution of video systems relative to photographic systems. This limitation can be circumvented with zoom lenses and auxiliary lenses that allow great enlargements of local zones, so that all the pixels of the video can be devoted to a small portion of the fringe pattern. Special considerations for this alternative involve laser power, accurate focusing, and accurate determinations of position and magnification.

In general, photographic films have sufficient resolution to record the whole-field fringe pattern in one photograph with a single magnification. This simplifies and shortens the course of the laboratory phase of the moire analysis. Subsequently, photographic enlargements can be made of any selected zone, to virtually any degree of magnification.

Extremely large fringe gradients are encountered in the analysis of composites. Consistent with the in-

troductory nature of this chapter, the photographic-film camera is recommended.

Low Light Levels

When modest laser power is used and larger image magnifications are demanded, the light coming from the ground-glass camera screen might be too weak for effective viewing. An easy modification can circumvent the problem: replace the ground-glass screen with a clear glass of the same thickness; adjust a focusing magnifier to focus on the inside surface of the clear glass; then, view the moire pattern through the focusing magnifier. The image will be much brighter, because all the light that reaches the screen is directed into the observer's eye. With a ground-glass screen, the light is spread in all directions and only a small portion reaches the eye.

Interpretation of Fringe Patterns

Fringe Ordering

The location of the zero-order fringe in a moire pattern is arbitrary. Any fringe—black, white or gray—can be assigned as the zero-order fringe. This is because rigid-body translations are not important in deformation analysis.

Adjacent fringes differ by plus- or minus-one fringe order, except in zones of local maxima or minima where they can have equal fringe orders. Local maxima and minima are usually distinguished by closed loops or intersecting fringes; in topography, such contours represent hills and saddle-like features, respectively. Fringes of unequal orders cannot intersect. The fringe order at any point is unique, independent of the path of the fringe count used to reach the point.

Fringe ordering is simple if the scheme proposed in the section *Sign of Fringe Gradient* is followed. Fringe orders merely increase (or decrease) monotonically until the point is reached where the sign of the gradient changes. From there, they decrease (or increase) monotonically.

Fringe ordering can often be rationalized without experimental determination of the signs of fringe gradients. The process involves advance knowledge of the sign of the gradient in some region, recognition of gradient changes at loops and saddles, and uniqueness. The process is risky, however, and requires experience.

Strain Analysis

Figure 9 provides an example of a thick graphite-epoxy $[90_2/0]_{11}$ specimen in interlaminar shear. In the N_x pattern, the zero-fringe order was assigned to the white fringe nearest the xy coordinate origin and fringe orders at the center of all other white fringes were numbered systematically as described above.

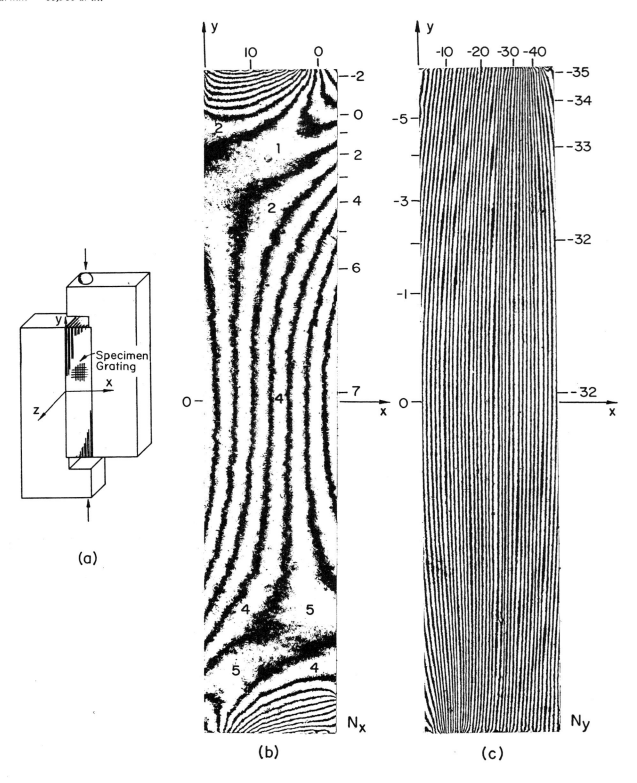

Fig. 9—Moire patterns for a thick graphite-epoxy specimen in interlaminar shear. (a) Rail-shear loading configuration. (b) N_x displacement field. (c) N_y field. Specimen length times width = 38 x 7.6 mm = 1.5 x 0.3 in. Reference grating frequency = 2400 l/mm = 60,960 l/in.

For N_y, the centers of black fringes were chosen as integral orders.

To determine ε_x along the horizontal centerline of the specimen, i.e., along the x axis, first plot the curve of N_x vs. Mx along the centerline, where M is the magnification of the image relative to the specimen. Then measure the slope of the curve at several points; the measured slopes are $\partial N_x/(M\partial x)$. From eq (3) the strain is

$$\varepsilon_x = \frac{M}{f}\left[\frac{\partial N_x}{M\partial x}\right] \qquad (8)$$

where the term in brackets is the measured slope. As an alternative, ε_x can be determined by the finite-increment relationship

$$\varepsilon_x = \frac{M}{f}\left[\frac{\Delta N_x}{M\Delta x}\right] \qquad (9)$$

To implement this, measure the distance $M\Delta x$ on the fringe pattern between two neighboring fringes. The change of fringe order between neighboring fringes is ± 1, and the term in brackets becomes $\pm[1/M\Delta x]$. The positive sign applies when N_x increases as x increases; otherwise the negative sign applies. Accordingly, eq (9) is implemented merely by measuring the x distance between fringes. Equation (9) is an approximation of eq (8), but where fringes are closely spaced or where the fringe gradient is not changing rapidly, it is a good approximation. It can be applied in a great many cases of moire-fringe analysis.

To determine ε_x along the vertical centerline, mark several stations along the centerline and measure $M\Delta x$ at these stations. The same method could be applied at the vertical boundaries, although the accuracy would be slightly lower. For better accuracy, plot N_x vs. Mx curves along several horizontal lines that intersect the vertical boundary; then find the slope of each curve at the boundary and use it in eq (8).

The same scheme can be applied to determine shear strain γ_{xy} at any point. The finite-increment relationship is

$$\gamma_{xy} = \frac{M}{f}\left[\frac{\Delta N_x}{M\Delta y} + \frac{\Delta N_y}{M\Delta x}\right] \qquad (10)$$

For the point of interest, measure distance $M\Delta y$ between neighboring fringes on the N_x pattern and measure $M\Delta x$ between neighboring fringes on the N_y pattern. The fringe-increment ΔN_x is $+1$ if N_x increases as y increases and otherwise it is -1; ΔN_y is $+1$ if N_y increases with x.

In regions where fringe gradients change very rapidly, special methods must be applied to extract the strain distribution in detail. Such cases arise in composites, especially for the analysis of interlaminar strains. In many such cases, the details can be extracted by adding carrier fringes to transform the load-induced fringes to a more tractable pattern. The method is explained in Ref. 5.

Digitizing Tablet—Computer Analysis

Measurements of the positions of integral-fringe orders can be facilitated by use of a digitizing tablet, a device that senses x,y positions and outputs the data through a personal computer. In use, an enlarged photographic print of the fringe pattern (e.g., N_x) is taped to the tablet. A focusing magnifier with a cross-hair reticle is placed on the photographic print. While viewing the pattern through the magnifier, the operator moves the cross-hair to any desired location, then presses a button to transmit the x and y coordinates of the point to the computer.

A digitizing tablet, computer and plotter can be employed to extract strains conveniently and effectively. Usually, an analysis follows these steps. The whole-field N_x and N_y fringe patterns are inspected to determine the region (or regions) of interest for strain analysis. This is usually the region of highest fringe gradients, where the strains are largest. An enlarged photographic print of the N_x (and/or N_y) pattern is made for that region. Lines are marked on the print where data are required and the print is taped to the digitizer tablet. The coordinates of the intersections of marked lines with fringe centerlines are sensed and recorded systematically by visual superposition of the cross hair with the intersections. Then, local strains can be determined from Δx (and/or Δy), using finite-increment eqs (9) and (10). Alternatively, data points of fringe orders vs. x or y can be plotted automatically. Curves can be fitted manually and slopes measured to implement eq (8) and its shear-strain counterpart. In principle, curves can be drawn and slopes can be extracted automatically, but caution is advised when investigating composites, because anomalous behavior and very steep gradients encountered in composites can sometimes be misinterpreted by curve-smoothing programs and real information can be filtered out in the process. Human intervention is encouraged for decision making to interpret the behavior of composites.

In addition to the advantages described above, the digitizing tablet approach offers that of human discrimination between information and noise in the fringe patterns. It appears to offer a very nice marriage of advantages when applied with moire interferometry to the analysis of composites.

Prospectus

Moire interferometry is relatively young and the techniques are advancing rapidly. We should anticipate further advances in laboratory practice, including advances in specimen-grating techniques and optical systems for generating the moire patterns. Efforts will be made to keep this chapter updated, but the investigator should also remain informed through the current literature of the mechanics and composites communities.

References

Publications

1. Czarnek, R., Post, D., and Guo, Y., "Strain Concentration Factors in Composite Tensile Members with Central Holes," Proc. 1987 SEM Spring Conf. on Experimental Mechanics, Houston, TX (June 14-19, 1987).

2. Post, D., Czarnek, R., Joh, D., Jo, J. and Guo, Y., "Deformation of a Metal-Matrix Tensile Coupon with a Central Slot: An Experimental Study," Journal of Composites Technology & Research, Vol. 9, No. 1, Spring 1987, pp. 3-9.

3. Post, D., "Moire Interferometry," Handbook of Experimental Mechanics, ed. A.S. Kobayashi, Ch. 7, Prentice-Hall, Englewood Cliffs, NJ (1987).

4. Post, D., Dai, F.L., Guo, Y. and Ifju, P., "Interlaminar Shear Moduli of Cross-Ply Laminates: An Experimental Analysis," J. Composite Materials, 23 (3), pp. 264-279 (March 1989).

5. Guo, Y., Post, D. and Czarnek, R., "The Magic of Carrier Fringes in Moire Interferometry," EXPERIMENTAL MECHANICS (to be published).

Vendors

Lasers
Spectra-Physics
Coherent
Newport

Mountings, Fixtures
Noll
Newport
Oriel

Photographic Plates
(Kodak HRP-type TE)
IMTEC

Optical Components
Edmund Scientific
Newport
Oriel
Physitec

Vacuum Deposition
Denton

Adhesives (replication)
Measurements Group

Addresses

Coherent, Inc.
Laser Products Div.
3210 Porter Drive
Palo Alto, CA 94303

Denton Vacuum Inc.
2 Pin Oak Ln.
Cherry Hill, NJ 08003

Edmund Scientific Corp.
7782 Edscorp Bldg.
Barrington, NJ 08007

IMTEC Products, Inc.
1295 Forgewood Avenue
Sunnydale, CA 94086

Measurements Group, Inc.
Micro-Measurements Div.
P.O. Box 27777
Raleigh, NC 27611

Newport Corp.
18235 Mt. Baldy Circle
Fountain Valley, CA 92708

J.A. Noll Co.
Box 312
Monroeville, PA 15148

Oriel Corp.
250 Long Beach Blvd.
Stratford, CT 05497

Physitec Corp.
206 Main St.
Norfolk, MA 02056

Spectra-Physics
Laser Products Div.
1250 W. Middlefield Rd.
Mountain View, CA 94039

Section IVB

Shearography: A New Strain-Measurement Technique and a Practical Approach to Nondestructive Testing

by Y.Y. Hung

Introduction

Because of their high strength to weight ratio, the usage of composite materials in load-resisting structures is increasing at a rapid rate. A composite is a combination of two or more materials and thus, the likelihood of flaws in composite materials is generally higher than that in metals. Consequently, there is a need to monitor the integrity of composite structures during and after fabrication. Since flaws and damage may develop during service, nondestructive inspections are also required in service. A nondestructive testing technique is needed not only to detect flaws but also to characterize flaws for the assessment of their severity and prescription of repair procedures. Until today, a practical method that can effectively and economically detect flaws in composite materials in a production/field environment has yet to be developed. Widespread implementation of nondestructive testing in production awaits the development of more reliable, and economical NDT techniques to meet the specific needs of various industries. The NDT method most extensively used at present is probably the ultrasonic technique. The ultrasonic technique, however, performs inspection on a point-by-point basis and therefore cannot meet the high inspection rate demanded by production. Another major limitation of the technique is that it generally requires fluid coupling, which is not practical for mass testing in a production/field environment.

This article presents an optical technique referred to as shearography for NDT of composites. Shearography measures derivatives of surface displacements and detects flaws in materials by looking for flaw-induced strain anomalies. It enjoys the advantages of being full-field, noncontacting, and having a high inspection speed; it also possesses a greater ability to identify the location and extent of a flaw. Shearography appears to have a great potential for being developed into a practical method for NDT of composites both in production and in service.

The Method

Shearography was originally developed for strain measurement[1]. It is an interferometric method allowing full-field measurement of surface-displacement derivatives. The object to be tested is illuminated by a point source of coherent light as shown in Fig. 1. It is imaged by an image-shearing camera equipped with a shearing device such as a glass wedge of small angle shown in Fig. 2. The image-shearing camera produces a pair of laterally sheared images in the image plane; hence, the method is named shearography. As the object is illuminated with coherent light, the two sheared images interfere with each other producing a random interference pattern commonly known as a speckle pattern. (Note that coherent light may be simply viewed as light that has the ability to interfere). When the object is deformed, this speckle pattern is slightly modified. Superposition of the two speckle patterns (deformed and undeformed) by double exposure yields a fringe pattern depicting the derivatives of the surface displacements. A flaw in the object usually induces a strain concentration which is translated into an anomaly in the fringe pattern.

Fig. 1—Schematic diagram of shearography

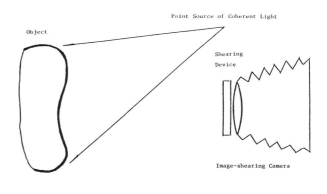

If observation of a real-time fringe pattern is desired, a photographic plate in the image plane of the camera is exposed with the object undeformed. The plate is processed in place or repositioned to its original position after processing. The object is then deformed. By viewing the image of the deformed object through the photographic plate of the unde-formed image, a real-time fringe pattern depicting the derivatives of the object displacement will be observed. This is a live fringe pattern which will change as the object is continuously deformed. The real time method can be used to monitor transient deformation.

Principle of Shearography

The Image-Shearing Camera

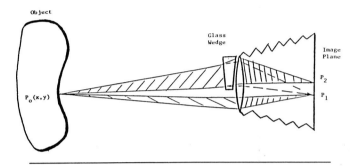

Fig. 2—The image-shearing camera

Shearography utilizes an image-shearing camera. This can simply be an ordinary camera equipped with an image-shearing device. While there are several means of producing a pair of sheared images in the image plane, a thin glass wedge illustrated in Fig. 2 is a simple but practical one. For best results, the wedge should be located in the iris plane of the lens covering one half of the lens aperture. Although a simple lens is illustrated, it is understood that a compound lens capable of producing an aberration-free image should be used. Without the wedge, rays scattered from an object point, say $P_o(x,y)$, and received by the two halves of the lens will converge to a point in the image plane; i.e., one point in the object is mapped into one point in the image plane. The glass wedge is a small angle prism which deviates rays passing through it. In the presence of the wedge, the rays from $P_o(x,y)$ in the object are mapped into two points P_1 and P_2 in the image plane. Thus, a pair of laterally sheared images is observed in the image plane. Rays are reversible. In other words, the image-shearing camera brings the rays scattered from one point on the object surface to meet with those scattered from a neighboring point in the image. If the wedge is so oriented that the rays are deviated in the xz plane, rays from a point $P(x,y)$ in the object will be brought to meet with the rays from a neighboring point in the x direction $P(x+\delta x,y)$. This is illustrated in Fig. 3.

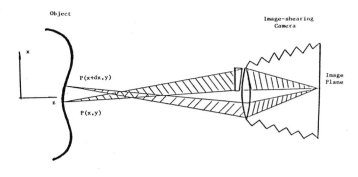

Fig. 3—Rays from two neighboring points are brought to meet in the image plane

δx, the amount of shearing in the object, is related to the wedge angle a by:

$$\delta x = D_o (\mu - 1)a \qquad (1)$$

where μ is the refractive index of the wedge material and D_o is the distance of the object from the wedge. Summing all the point pairs results in a pair of images laterally sheared in the x direction.

Analysis of Fringe Formation

Since the object is coherently illuminated, the two images interfere with each other producing a result-ant image intensity distribution I given by[1]*:

$$I = I_o (1 + \cos \phi) \qquad (2)$$

where I_o is the object image and ϕ is a random phase angle. Thus the object image is modulated by $(1 + \cos \phi)$ which represents a random interference pattern commonly referred to as speckle pattern.

When the object is deformed, an optical path change occurs due to the surface displacement in the object. This optical path change produces a relative phase change between the two interfering points. Thus, the intensity distribution of the speckle pattern is slightly altered and is mathematically represented by:

$$I' = I_o [1 + \cos (\phi + \Delta)] \qquad (3)$$

where I' is the intensity distribution after deformation, and Δ is the relative phase change due to relative displacement between $P(x,y)$ and $P(x+\delta x,y)$.

A photographic film is doubly exposed sequentially to I and I'. The total energy recorded is proportional to the sum I_s of the two intensity distributions:

$$I_s = I + I' = I_o[2 + \cos \phi + \cos (\phi + \Delta)] \qquad (4)$$

* Footnotes refer to references

ϕ in the above equation has a random value and is fast varying; thus the spatial frequencies of cos ϕ and cos (ϕ + Δ) are generally beyond the eye's resolution capability. Hence the eye sees an intensity averaged over an elementary area resolvable by the eye. Since ϕ is random, the probability of ϕ having any value between 0 and 2π is equal. Therefore, the averaged intensity is shown[1] to be equal to $2I_o$. This means no fringe pattern is seen. A weak fringe pattern may be observed if the response of the recording emulsion is nonlinear[2]. However, a highly visible fringe pattern can be obtained by a high-pass spatial frequency filtering process to be described later.

Equation (4) may be rewritten in the following form:

$$I_s = 2I_o[1 + \cos(\phi + \frac{\Delta}{2}) \cdot \cos(\frac{\Delta}{2})] \quad (5)$$

The second term in the above equation represents a high-frequency carrier cos $(\phi + \frac{\Delta}{2})$ modulated by a low-frequency factor cos $(\frac{\Delta}{2})$. The high-frequency carrier is nullified when:

$$\cos(\frac{\Delta}{2}) = 0 \quad (6)$$

The fringe pattern of eq (5) is classified as frequency-variation type. Fringes of this type are identified as areas where the carrier is absent as described by the condition given in eq. (6). It should be noted this type of fringe pattern is different from a conventional fringe pattern. Conventional fringes are amplitude modulated and are readily visible. Fringe lines of a conventional fringe pattern are loci of points at which intensity is minimum or maximum. The present frequency-variation type of fringe pattern is frequency modulated and is not readily visible. Fringe lines are loci of zero frequency, i.e. absence of speckles. However, these invisible fringes can be converted to a visible intensity-variation type by an optical high-pass Fourier filtering process. The fringe pattern depicts Δ, the relative surface displacement between two neighboring points.

Fringe Readout by Fourier Filtering

Fig. 4 shows a schematic diagram of the Fourier filtering setup. A point source of light is focused by the transforming lens (TL) onto a plane known as the Fourier filtering plane. The photographic transparency with an intensity transmittance represented by eq (5) is placed at the input plane. The light field appearing in the Fourier filtering plane is the Fourier transformation of the transmittance. Fourier transform is a transformation from a spatial domain to a frequency domain. In other words, carriers of various frequencies recorded in the transparency are separated in the Fourier filtering plane. The areas of zero frequency are displayed at the focal point on the

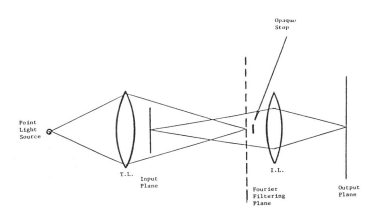

Fig. 4—Schematic diagram of high-pass Fourier filtering

optical axis, whereas those of higher frequencies are located off-axis in the frequency plane. By placing an opaque stop on the axis at the Fourier filtering plane to block the zero- and low-frequency components, the contribution from the areas (fringe lines) where the carriers are nullified is stopped. The process is known as high-pass spatial filtering. When reimaged by the imaging lens (TL) behind the filtering plane, those areas will appear dark in the image plane (output plane). Thus, a frequency-variation-type fringe pattern is converted into a visible fringe pattern of intensity variation. Since the useful information is coded in the interference pattern obtained by the light imaged by the two halves of the recording lens of the image-shearing camera, this speckle pattern has a major spatial frequency component in a direction perpendicular to the diameter dividing the two halves and parallel to the plane of the lens. Optimal filtering can be achieved by placing a filtering aperture off-axis in the frequency plane to receive this frequency component. It is recommended to use a white-light source for the filtering to minimize the speckles in the output image.

Fringe Interpretation

With the Fourier filtering, an invisible frequency-variation type of fringe pattern is converted into a visible intensity-variation fringe pattern depicting Δ, the relative phase change due to the relative displacement between two interfering points, $P(x,y)$ and $P(x + \delta x, y)$. Fringe lines corresponding to the areas of nullifying carrier become dark fringe lines after filtering. The fringe orders are related to Δ by eq (6):

$$\Delta = (2n + 1)\pi \quad (7)$$

where $\pm n = 0, 1, 2, 3, 4, 5$, are fringe orders of dark fringe lines.

Δ is related to the relative displacements by[1]:

$$\Delta = \frac{2\pi}{\lambda} (A\,\delta u + B\,\delta v + C\,\delta w) \qquad (8)$$

where (u, v, w) and $(u + \delta u,\ v + \delta v,\ w + \delta w)$ are the displacement vectors of $P(x, y)$ and $P(x + \delta x, y)$, respectively. A, B and C are sensitivity factors related to the position of the illumination point $S(x_s, y_s, z_s)$ and the camera position $O(x_o, y_o, z_o)$ by

$$A = \frac{(x - x_o}{R_o} + \frac{x - x_s)}{R_s}$$

$$B = \frac{(y - y_o}{R_o} + \frac{y - y_s)}{R_s} \qquad (9)$$

$$C = \frac{(z - z_o}{R_o} + \frac{z - z_s)}{R_s}$$

where $R_o^2 = x_o^2 + y_o^2 + z_o^2$ and $R_s^2 = x_s^2 + y_s^2 + z_s^2$. Note that $z = z(x, y)$ describes the object surface.

Equation (8) may be rewritten as

$$\Delta = \frac{2\pi}{\lambda} \left(A\frac{\delta u}{dx} + B\frac{\delta v}{dx} + C\frac{\delta w}{dx} \right) dx \qquad (10)$$

If the amount of shearing, dx, is small, the relative displacements approximate the derivatives of displacements. Thus,

$$\Delta = \frac{2\pi}{\lambda} \left(A\frac{\partial u}{\partial x} + B\frac{\partial v}{\partial x} + C\frac{\partial w}{\partial x} \right) dx \qquad (11)$$

By rotating the lens-wedge assembly 90 deg about the optical axis, the shearing direction becomes parallel to the y-direction, and eq (11) becomes:

$$\Delta = \frac{2\pi}{\lambda} \left(A\frac{\partial u}{\partial y} + B\frac{\partial v}{\partial y} + C\frac{\partial w}{\partial y} \right) \delta y \qquad (12)$$

Equation (12) describes the displacement derivatives with respect to y, where δy is the amount of shearing in the y direction. It is possible to employ a multiple image-shearing camera[3]—permitting displacement derivatives with respect to both x and y to be recorded simultaneously.

Strain Measurement

Deduction of Strain

Equations (11) and (12) indicate that the obtained fringes are generally functions of three displacement derivatives, namely,

$$\frac{\partial u}{\partial x}, \quad \frac{\partial v}{\partial x} \quad \text{and} \quad \frac{\partial w}{\partial x} \quad \text{or} \quad \frac{\partial u}{\partial y}, \quad \frac{\partial v}{\partial y} \quad \text{and} \quad \frac{\partial w}{\partial y}.$$

Therefore, three measurements with three different illuminating angles are required for the three displacement derivatives to be separated. These displacement derivatives allow surface strains to be determined as:

$$\varepsilon_x = \frac{\partial u}{\partial x}$$

$$\varepsilon_y = \frac{\partial v}{\partial y} \qquad (13)$$

$$\gamma_{xy} = \frac{\partial v}{\partial x} + \frac{\partial u}{\partial y}$$

The technique is particularly well suited to the measurement of flexural strains in plate structures, as the flexural strains are related to change of surface curvatures. Flexural strains are related to the second derivatives of plate deflections by:

$$\varepsilon_x = \frac{\partial^2 w}{\partial x^2}$$

$$\varepsilon_y = \frac{\partial^2 w}{\partial y^2} \qquad (14)$$

$$\gamma_{xy} = 2\frac{\partial^2 w}{\partial x\,\partial y}$$

For measurement of flexural strain in plates, it is desirable to employ normal illumination and normal viewing, that is, $x_s = y_s = x_o = y_o = 0$, and for R_o and R_s to be large compared with the object size; then eqs (11) and (12) are reduced to:

$$\Delta = \frac{4\pi}{\lambda} \left(\frac{\partial w}{\partial x} \right) \partial x \qquad (15)$$

$$\Delta = \frac{4\pi}{\lambda} \left(\frac{\partial w}{\partial y} \right) \partial y \qquad (16)$$

Thus only the derivatives of w are needed; these can be achieved by one recording. Figure 5(a) shows a fringe pattern depicting $\frac{\partial w}{\partial x}$ of the deformation in the footprint of a tire. The fringe pattern of Fig. 5(b) depicts $\frac{\partial w}{\partial y}$ of the footprint deformation. The obtained data are then differentiated once to yield strains. However, numerical differentiations are prone to large errors. Direct measurement of flexural strains is possible by employing the shearographic technique reported in Ref. 4.

Fig. 5—Deformation in the footprint of a tire.

(a) $\frac{\partial w}{\partial x}$ and (b) $\frac{\partial w}{\partial y}$

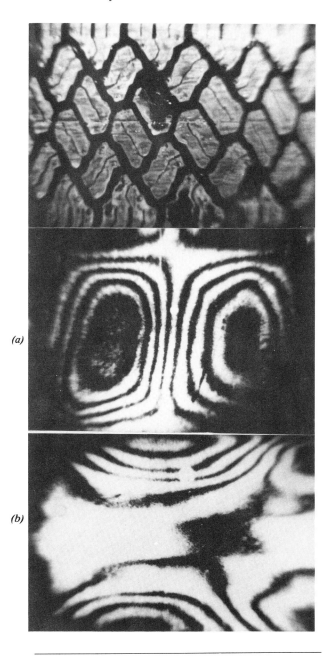

(a)

(b)

Computerization of Data Deduction

Obviously, one major difficulty in employing shearography for strain measurement is the identification of fringe orders. Like other optical techniques in which the output is in the form of fringe patterns, there are ambiguities in determining fringe orders and their signs. This is because order 1 looks

like order 2 or order -2 or any other fringe order. At present, one common practice to identify fringe orders relies upon prior knowledge of the problem being investigated, such as boundary conditions and the nature of the problem.

The fringe ambiguities can be removed by a recently reported carrier-fringe technique[5]. The technique is based on the introduction of known linear carrier fringes which are coherently superimposed upon the deformation fringes. The carrier fringes make the fringe orders of the resulting fringe pattern monotonically increasing/decreasing, and thus can be determined without ambiguity. The phase change due to deformation alone is obtained by subtracting the known linear phase variation from the combined fringe pattern.

Fig. 6—Fringe pattern depicting $\frac{\partial w}{\partial x}$ of a rectangular plate

clamped along its edges and subjected to a uniform pressure

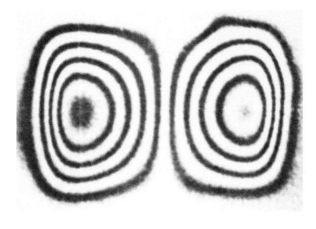

The following example will be used to explain the principle. Figure 6 shows a fringe pattern depicting $\frac{\partial w}{\partial x}$ of a rectangular plate clamped along its four edges and subjected to a uniform pressure. Coherent superposition of linear carrier fringes results in a deformed carrier-fringe pattern shown in Fig. 7. The order of the carrier fringes of Fig. 7 increases monotonically from left to right. A plot of the fringe order along the horizontal center line is shown in Fig. 8. The straight line in Fig. 8 is the plot of the reference fringes (i.e., carrier fringes without deformation). Subtraction of the reference-fringe orders from the deformed carrier-fringe orders produces the distribution of $\frac{\partial w}{\partial x}$ along the center line of the plate which is shown in Fig. 9. A comparison with the predicted results is also shown. The data deduction process is carried out in a straight forward manner

Fig. 7—Deformed carrier fringe pattern produced by coherent superposition of a linear phase variation upon the deformation fringes of Fig. 6.

Fig. 7—Deformed carrier fringe pattern produced by coherent superposition of a linear phase variation upon the deformation fringes of Fig. 6.

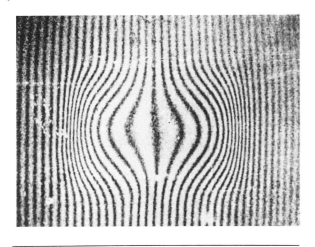

Fig. 8—Plots along the center lines of the plate for the deformed and undeformed (straight line) carrier fringes

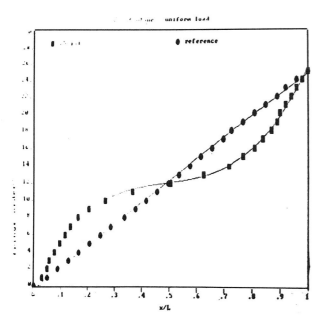

Fig. 9—The distribution of $\frac{\partial w}{\partial x}$ by the straight forward subtraction

of the undeformed carrier fringe order from that of the deformed

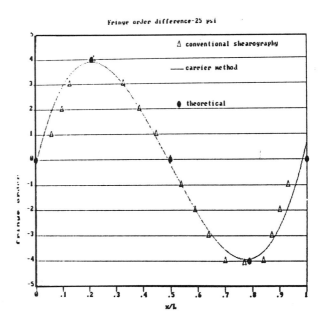

and without ambiguity. This carrier-fringe technique has paved a way for automating data deduction. With this technique, computerization of the data deduction has been demonstrated by means of digital image processing reported in Ref. 6.

Nondestructive Evaluation and Methods of Testing

Flaw Detection

The first question one may ask is: "How does shearography detect flaws in materials?" The answer is: "Shearography reveals flaws by looking for flaw-induced strain anomalies which are translated into anomalies in the fringe pattern." Although shearography measures surface deformation, it can detect both surface and internal flaws. This is because the internal flaws, unless very remote from the surface, also affect the surface deformation. The size and location of a flaw are directly determined by the size and location, respectively, of the fringe anomaly, and the nature of the flaw can be determined from the signature of the fringe pattern.

Although shearography was originally developed for strain measurements, it is more readily accepted by industry for nondestructive testing than strain measurement. This is due to the fact that identification of fringe orders is generally not required in NDT applications. Despite its youthfulness, shearography has already been proven to be a valuable nondestructive inspection tool for the rubber industry. It is now used routinely to evaluate the integrity of tires, particularly, aircraft tires. FAA has endorsed shearography as a standard method for inspecting aircraft tires. Other successful applications of shearography in nondestructive testing include evaluation of pressure vessels, boiler tubes, adhesive bonds, grinding wheels, riveted joints, concrete structures, integrity of welded joints, and detection of leakage in the seal of computer IC packages. An earlier report of these applications can be found in Ref. 7.

Since tires are made of composite laminates, the applicability of shearography for NDT of composites has already been demonstrated. Investigation of shearography for detection of delaminations in composition laminates was also reported[8]. Thus far, it has been demonstrated that shearography can effectively detect flaws such as delaminations, broken fibers, matrix cracking and presence of moisture. This report will focus primarily on NDT of composites.

Method of Stressing

Flaw revealment by shearography is based on the comparison of two states of deformation in the test object. Development of NDT procedures employing shearography essentially becomes the development of a practical means of stressing which can reveal flaws. Ideally, it is desirable to impose stresses identical to the stress state found in service. If components under testing are loaded in a stress mode similar to the actual one experienced in service, shearography can be used to reveal critical flaws only (i.e., flaws that create strain concentrations and thus reduce the strength of the component). Cosmetic flaws can be ignored and false rejects can be avoided. Examples of cosmetic flaws include those located in low-stress regions which will not jeopardize the strength of the structures. In this regard shearography has an advantage over ultrasonic techniques. Ultrasonic techniques detect flaws by identifying inhomogeneities in the materials and provide no information about the criticality of the flaws.

However, exact duplication and application of actual loading during testing may be difficult or impractical. Five practical methods of stressing for testing composites are described below. One precaution in stressing the structures is the prevention of rigid-body motion. Excessive rigid-body motion would cause decorrelation of the speckles in the two images (deformed and undeformed) resulting in degradation of fringe quality.

Pressurization

This stressing mode is ideal for pressure vessels and those structures that can be pressurized. With the double-exposure technique, the structure under test is pressurized between the exposures. The sequence of the exposures is immaterial. The structure may also be initially pressurized and then additionally pressurized. Pressurization usually does not introduce intolerable rigid-body motion. Since internal pressurization represents the actual stressing for pressure vessels and pipes, the criticality of the flaws revealed can be determined readily from the level of strain concentration.

Internal pressurization may also be employed to inspect honeycomb structures. Honeycombs are generally sealed and therefore can be pressurized. By drilling a small hole through the skin, the structure can be internally pressurized through the hole. The hole is then mended after testing. Of course the hole chosen should be located in a less critical region. This technique, if accepted, can be developed into a method for inspecting honeycombs in an aircraft in the field. A permanent access of pressurization may be built in the structures for regular inspection in service.

Partial Vacuum Stressing

This is the stressing mode used in the testing of tires. It is also an effective stressing technique for revealing delaminations in composite laminates and honeycomb type of structures. In the testing, the object and the optics are placed in a chamber where a partial vacuum can be drawn. Partial vacuum is applied between the double exposures. Partial vacuum is equivalent to a uniform tensile force applied to the object surface which pulls the surface outward. Thus a debond or delamination causes the surface directly above the flaw to bulge out slightly, which can be detected with shearography. Presence of trapped air in the delaminated region further aids in the revealment. In the testing of tires and inspection of skin-to-core bonding in honeycombs, a partial vacuum in the order of one psi is generally sufficient.

For large objects of which the total enclosure is not practical, application of partial vacuum to a small area at a time can be performed with a suction cup having a transparent window. A schematic of the device is shown in Fig. 10. The suction cup is first sealed against the test surface by applying an initial vacuum. The test area is illuminated and imaged by an image-shearing camera through the transparent window. First exposure is made with the initial vacuum applied, and the second exposure after application of additional vacuum. Deformation of the window may distort an additional phase change in the fringe pattern but is generally tolerable. This technique is applicable to inspection of aircrafts in the field.

Thermal Stressing

In this type of testing, a double-exposure shearograph is made with the object surface being radiated with heat between the exposures. The temperature gradient developed induces thermal stresses in the ob-

87

Fig. 10— *Schematic of the suction cup used to apply vacuum stressing*

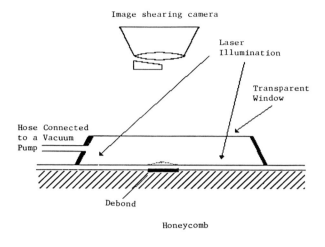

Vibrational Excitation

ject. This stressing mode is particularly well suited to the evaluation of bonding between two different materials. The difference in the coefficient of thermal expansion between the materials gives rise to a quasi-bimetallic strip. The debonded area is not constrained and therefore is free to deform away from the interface; this in turn produces a strain anomaly on the surface. In the case where there is trapped air in the debonded region, the heat will cause the trapped air to expand, causing the material above the flaw to bulge out. Usually a steady-state thermal deformation may not be easily maintained during the exposure time. In this case, real-time shearography should be employed to observe the transient thermal deformation.

Vibrational Excitation

Vibrational excitation is another effective means for revealing delaminations. This is a dynamic stressing method. In the testing, the structure is excited by a transducer such as a piezoelectric crystal which induces an acoustical wave to propagate into the structure. A real-time technique should be used in the testing to observe the vibration modes of the debonded area. The debonded area will vibrate in a set of normal modes as the scanned frequency of the excitation coincides with the resonant frequency of the debonded area. The fundamental resonance frequency of the debond depends on the material properties and it is proportional to the depth of the debond and inversely proportional to the square of the radius of the debond. This theory assumes the debond behaves like an isotropic circular plate clamped along its edges.

Microwave Excitation

Microwave stressing is used to detect the presence of moisture in materials. Between the exposures, the object is excited by microwave having the frequency which excites water molecules. A microwave gun from a home microwave oven would serve the purpose. The microwave excitation causes the moisture in the materials to heat up and thus induces highly localized deformation detectable by shearography. This mode of stressing is only applicable to nonmetallic composites.

Flaw Characterization

Recent development in fracture mechanics and structural-life management demand that nondestructive testing techniques not only detect, but also characterize flaws. Flaw characterization is needed to determine flaw criticality and predict the remaining life of components. Criticality of a flaw depends upon flaw size, its shape and location, as well as the nature of the flaw. Shearography allows flaw location and flaw size to be assessed readily. Analysis of the flaw nature is more complicated as it generally requires a comparison of the signature of the fringe anomaly with a database. Much work is needed to advance the state-of-the-art of flaw characterization with shearography. So far, only debonds can be characterized with confidence.

A simple model used to describe a debond in a laminate composite is a circular plate clamped along its edges. When partial vacuum is used to reveal the debond, the plate bulges out. The displacement of the plate surface is illustrated in Fig. 11(a), and a plot of the plate deflection W is shown in Fig. 11(b). The contour lines of the plate deflection appears in the form of a bull's eye. This is a fringe pattern one would expect when the component is inspected with holography. Note that holography measures surface displacements. However, shearography measures derivatives of the displacement. The derivatives of the plate deflection with respect to x are shown in Fig. 11(c). The contour map of the deflection derivative is in the form of a double bull's eye. Thus, in shearography, a debond is characterized by a double bull's eye fringe pattern as illustrated in Fig. 12. For a rectangular-shaped debond of size $a \times b$, a double bull's eye fringe pattern of size $(a+s) \times b$ will be observed, where s is the amount of shearing in the a direction. Thus for a double bull's eye of size $a' \times b$, the actual size of the debond is estimated to be $(a'-s) \times b$. This is illustrated in Fig. 13.

The depth of a debond can be estimated from the fringe density. For debonds of the same size, the one closer to the surface has a higher fringe density and vice versa. This is illustrated in Fig. 14, which shows four separations of approximately same size but located at depths of 3, 6, 9, and 12 mm from the sur-

Fig. 11—*Deformation of a typical debond subjected to vacuum stressing*

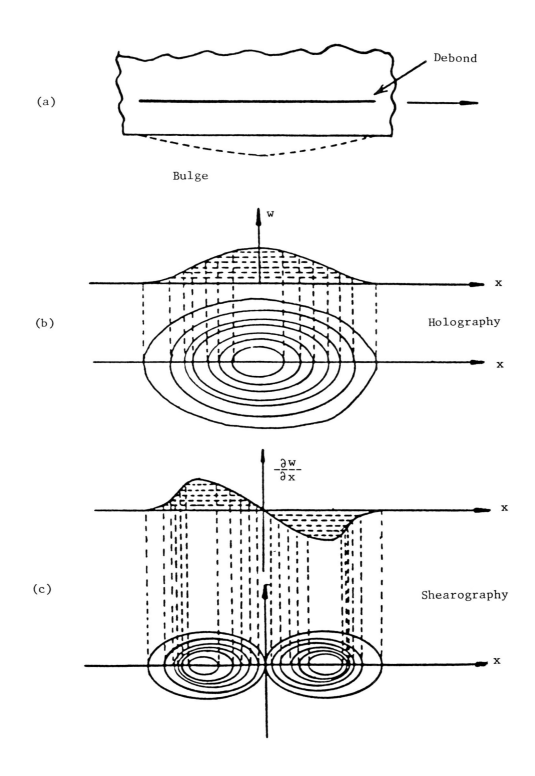

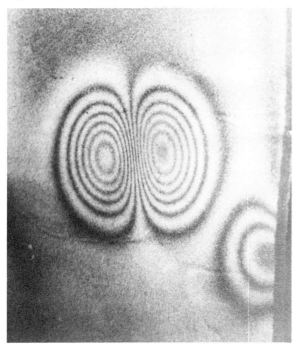

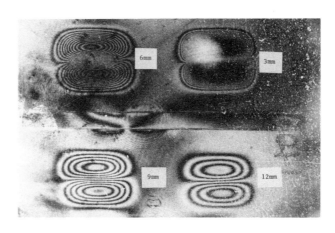

face. The size of the separations can be estimated by the areas of the butterfly fringe pattern, and the depth from the fringe density. Notice that the one closest to the surface has the highest fringe density and vice versa. This is because a flaw closer to the surface has greater influence on the surface deforma-tion. The plate theory predicts that the deflection is inversely proportional to the flexural rigidity of the plate, and the flexural rigidity is proportional to the cube of the thickness. This means for equal size debonds, the density of the fringes is inversely pro-portional to the cube of the depth. This theory is for plates of isotropic materials.

Demonstration

Delamination in a composite cylinder. The double bull's eye fringe pattern in Fig. 15 reveals a delamina-tion in a filament-wound cylinder. The means of stressing used is internal pressurization.

Fig. 13—The sheared images of a debond of size a x b, s is the amount of shearing

Fig. 15—Fringe pattern revealing a delamination in a filament wound cylinder

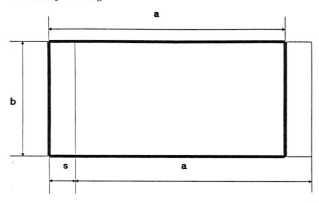

Delaminations in a composite honey panel. Five programmed delaminations in a honeycomb panel were revealed by the shearographic fringe pattern shown in Fig. 16. The skin of the honeycomb is made of carbon epoxy laminates. The means of stressing is partial vacuum.

Fig. 16—Programmed and unexpected delaminations in a honeycomb panel

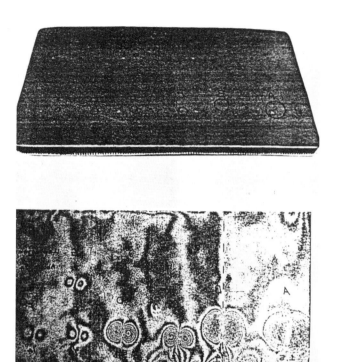

Fig. 18—Shearographic fringe pattern revealing unbonds at the interface of a plastic/metal bond

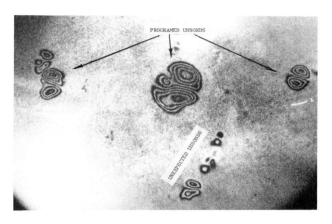

Flaws in a cord-reinforced rubber panel. Figure 19 shows various delaminations and a long internal crack in a cord-reinforced rubber panel. The method of stressing used is partial vacuum.

Fig. 19—Flaws in a cord-reinforced rubber panel, including a vertical internal crack

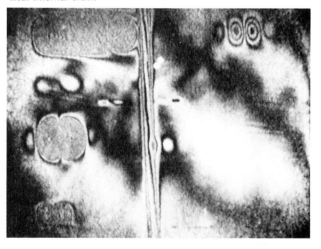

Flaws in a tire. Figure 17 shows a fringe pattern revealing separations along the belt edge of a truck tire. Note that at the steel-belt edge region, there is an abrupt transition in material stiffness which tends to induce separations in tires. The means of stressing is partial vacuum.

Flaws in an adhesive bond. Figure 18 shows a fringe pattern revealing three programmed debonds at the interface of a plastic/metal bond. Although the sample was carefully prepared in a laboratory, unexpected debonds were detected.

Fig. 17—Separation along the belt edge of a steel-belted tire

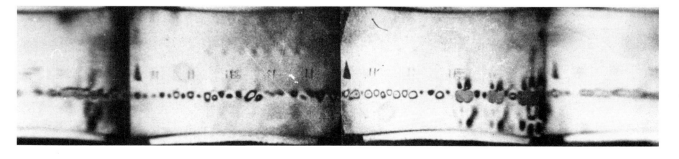

Fig. 20—(a) The steel-reinforced concrete sample; (b) A broken reinforcement rod produces a fringe signature of triangular shape

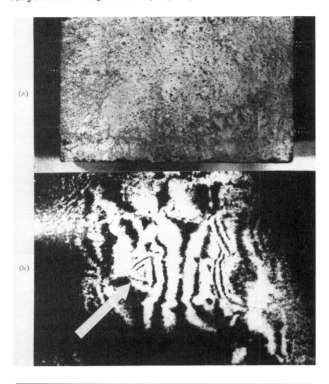

Steel-reinforced concrete sample. The triangle-shaped fringe pattern of Figure 20 reveals a broken steel rod in a steel reinforced concrete slab. The means of stressing is heating.

Fig. 21—The abrupt change in the fringe direction reveals a crack in a composite blade

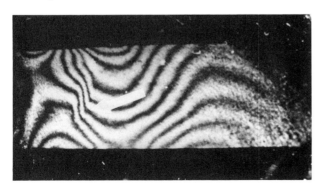

Crack in a composite blade. The fringe pattern of Figure 21 shows the presence of a crack in a graphite-epoxy blade. The crack is characterized by the abrupt change in the fringe lines. Heating was used to stress the blade.

Shearography vs. Holography

Both shearography and holography are optical methods which reveal imperfections in materials by identifying defect-induced deformation anomalies. A study was performed to assess the capability of each technique in the evaluation of tires and composites. Figure 22 compares the results for detecting interfacial debonds. The means of stressing in both cases was vacuum. The superiority of shearography over holography in this application is clearly demonstrated. The fringe pattern (Fig. 22a) of shearography reveals the disbond more prominently than that (Fig. 22b) of holography. This is because shearography measures strains whereas holography measures displacements. Since defects normally create strain concentrations, it is easier to correlate defects with strain anomalies rather than displacement anomalies. Particularly, shearography, which measures displacement gradients, is not sensitive to rigid-body motions. Note that rigid-body motion produces confusing, if not misleading, fringes in holography.

Other advantages of shearography over holography are summarized as follows. (a) It requires a very simple optical setup, thus eliminating the optical alignment problems. (b) It relaxes the vibration-isolation requirements; thus, it is more suited for inspection in field/production environments. (c) The coherent length requirement of the light source is greatly reduced, thus eliminating the problem of maintaining lasers running at single mode. (d) It provides a wider and more controllable range of sensitivity for many practical measurements. One additional control of sensitivity in shearography is by varying the magnitude of shearing. (e) The requirement of recording media resolution is much lower, thus allowing faster and less expensive photographic films to be used; even video recording is possible. (f) Fringe interpretation is simpler, thus offering the possibility of automatic fringe deduction by means of computer image processing. Ths is primarily due to the absence of rigid-body-motion, fringe pattern.

Computer-Aided Shearography

Shearography has not been widely used yet. One of the reasons is the need to use photographic film as the recording medium, which is slow and costly. The subsequent Fourier filtering process needed for the readout of fringe patterns further delays the output of the testing results. Furthermore, human reading

Fig. 22—Shearographic fringe pattern (a) versus holographic fringe pattern (b) of a debond

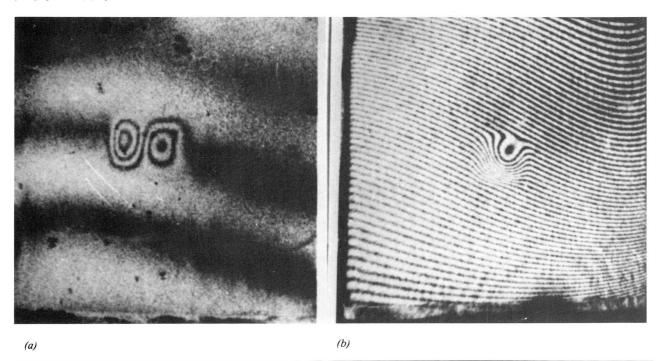

(a)   (b)

and interpretation of fringe patterns can be subjective and unreliable. Research has been devoted to automate the testing process. Recently, a breakthrough has been made which eliminates the photographic recording and the subsequent fringe readout process. This new technology is referred to as electronic shearography. Electronic shearography employs a video camera as a recording medium and digital image processing is used to process and form the fringe pattern. With the aid of computers, shearographic nondestructive testing can be performed at video rate. The author will report the technique in a future paper.

Some examples of the results obtained by electronic shearography are presented here. Figure 23 shows a fringe pattern obtained by electronic shearography depicting $\frac{\partial w}{\partial x}$ of a rectangular plate clamped along its four edges and centrally loaded. The butterfly fringe patterns of Figure 24 are indicative of delaminations in a composite laminate.

Fig. 24—Butterfly fringe pattern of electronic shearography revealing two delaminations in graphite epoxy laminates

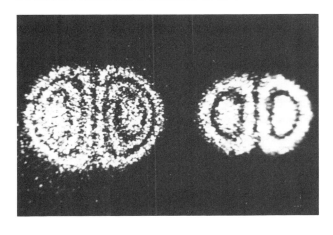

Fig. 23—Fringe pattern of electronic shearography depicting $\frac{\partial w}{\partial x}$ of pressure loaded rectangular plate clamped along its edges

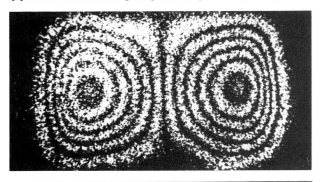

The fringe pattern shown in Figure 25 reveals a damaged region in a plate of graphite-epoxy laminates due to impact. Figure 26 is a comparison of electronic shearography with the result obtained by a c-scan ultrasonic technique in the test of a graphite panel with rohacell core. The edge pullout and a teflon insert are detected by both techniques. However, electronic shearography revealed the flaws in a fraction of a second, whereas ultrasonic requires point-by-point scanning of the part. Moreover, fluid couple is needed in the ultrasonic testing.

Fig. 25—Impact damages in a plate of graphite epoxy detected by electronic shearography

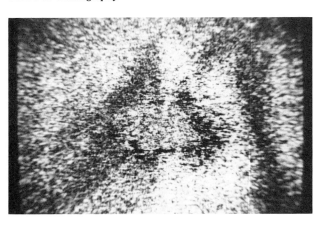

Fig. 26—Comparison of electronic shearography with c-scan ultrasonic technique. The edge pullout and a teflon insert in a graphite panel are detected by both techniques. (Photograph courtesy of Laser Technology, Inc.)

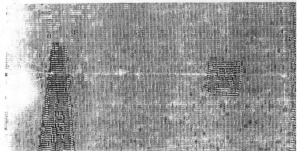

Fig. 27

(a) Three delaminations in a composite panel

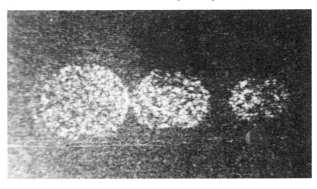

(b) Automatic identification by a digital image processing technique

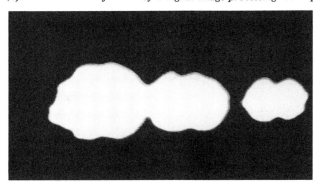

Some progress has also been made in the area of automated flaw identification. Digital image processing techniques are being developed by the author to automatically identify fringe anomalies. An example of some preliminary results is shown in Figure 27. Figure 27(a) shows three delaminations in a composite laminate revealed by electronic shearography. Figure 27(b) is the result of a digital image-processing technique which automatically identifies fringe anomalies and outlines the size and location of the flaws.

Conclusion

Shearography represents a new approach to nondestructive testing. The viability of the shearographic nondestructive testing of a given structure lies primarily in the stressing technique. Given a particular structure to be tested, one must determine an effective means of stressing which will best reveal flaws in the structure. Shearography employing vacuum stressing has been proven to be a very effective technique for revealing delaminations in composites. A delamination revealed by partial vacuum is characterized by a fringe pattern in the form of a double bull's eye. The fringe anomaly readily indicates the flaw size and location.

Sheography is a practical NDT technique which is rapidly gaining acceptance by industry. The development of the electronic version has greatly facilitated the employment of shearography as an NDT tool in production and in service. It has been successfully employed to inspect composite structures in a production as well as a field environment. Shearography is relatively young; its full capability awaits further exploration.

Acknowledgment

The author is grateful to Mrs. Mary Sue Perria for her help in preparing this manuscript. Her patience and dedication are greatly appreciated.

The research reported in this article was supported, in part, by the National Science Foundation (Grant No. MSM-8815420). The support of Dr. Oscar Dillon is appreciated.

References

1. Hung, Y.Y., "Shearography: A New Optical Method for Strain Measurement and Nondestructive Testing," Opt. Eng., 391-395 (May/June, 1982).

2. Hung, Y.Y., "A Speckle-Shearing Interferometer," Optics Commun., 11, 732 (1974).

3. Hung, Y.Y. and Durelli, A.J., "Simultaneous Measurement of Three Displacement-Derivatives Using a Multiple Image Shearing Interferometric Camera," J. Strain Anal., 14 (3), 81-88 (1979).

4. Takezaki, J., and Hung, Y.Y., "Direct Measurement of Flexural Strains in Plates by Shearography," J. Appl. Mech., 53, 125-129 (March 1986).

5. Hung, Y.Y., Hovanesian, J.D., and Takezaki, J., "A Fringe Carrier Technique for Unambiguous Determination of Fringe Orders in Shearography," Optics and Lasers in Engineering, 8 (2), 73-81 (1988).

6. Templeton, D.W., and Hung, Y.Y., "Computerization of Data Deduction in Shearography," Proc. 31st Ann. Int. Techn. Symp. on Optics and Optoelectronic Appl. Sci. and Eng., San Diego, CA, Aug. 16-21, 1987.

7. Hung, Y.Y. and Hovanesian, J.D., "Shearography: A New Nondestructive Testing Method," Use of New Technology to Improve Mechanical Rediness, Reliability and Maintainability, Proc. 40th Mtg. of the Mechanical Failures Prevention Group, NBS, Gaithersburg, MD, April 16-18, 1985.

8. Anastasi, R.F., Serabian, S.M., Shuford, R.J., and DasGupta, D.K., "Nondestructive Detection of Simulated Delaminations in Composite Laminates by Laser-Speckle Shearography," EXPERIMENTAL TECHNIQUES, 28-31 (June 1987).

Section VA

Acoustic Methods of Evaluating Elastic Properties
or,
Will the Real Young's Modulus Please Stand Up?

by V.K. Kinra and V. Dayal

ABSTRACT—Measurement of the elastic moduli using ultra-sound has become fairly routine. One measures a speed of sound, c, the density, ρ, and calculates a modulus, E, from a formula of the type $c^2 = E/\rho$. Now, E describes a *static* response whereas c describes a *dynamic* response of the material. The connection between the two, $E = \rho c^2$, is strictly valid for an ideally elastic and homogeneous material; it remains valid for heterogeneous materials, e.g. composites, so long as a key assumption is satisfied, namely, the wavelength is large compared to any characteristic length of the material. In metals, a characteristic length is the grain size and generally the wavelength is large compared to the grain size. By their very definition, composites are *heterogeneous* materials and have one or more characteristic lengths. The objective of this paper is to demonstrate that when the wavelength becomes of the order of the characteristic length of a composite, large errors may occur and the *dynamic* measurement of E may differ from its true static value by as much as 200 percent; hence the subtitle: 'Will the real Young's modulus please stand up?'

List of Symbols

E = Young's modulus, GPa

V_w = velocity of sound in water, mm/μs

a = inclusion radius, mm

c_1 = longitudinal wavespeed, mm/μs

c_2 = shear wavespeed, mm/μs

$\tilde{c}$ = volume fraction

d = a characteristic length, mm

k = wavenumber, mm^{-1}

n = frequency, cycles/s

t = travel time, s

w = specimen thickness, mm

Ω = normalized frequency, $k_1 a$

λ, μ = Lame's constants

λ = wavelength, mm

ν = Poisson's ratio

ρ = specific gravity

ω = frequency, radians/s

$<>$ = aggregate property of composite

Introduction

We have measured the velocity of longitudinal and shear waves, $<c_1>$ and $<c_2>$ respectively, in random and periodic particulate composites. The inclusions were spheres of a single size; therefore, the only characteristic length of the composite is the radius a. Let λ_1 be the wavelength of the longitudinal wave in the matrix material, the wave number $k_1 = 2\pi/\lambda_1$, and introduce a normalized frequency $k_1 a = 2\pi n a/c_1$, where n is the frequency in cycles/time; recall that $c_1 = \omega/k_1$ where $\omega = 2\pi n$. To guard against fortuitous or material-specific results, we have tested four disparate material systems: (1) inclusion which is very heavy but not very stiff as compared to the matrix: lead/epoxy; (2) inclusion which is very stiff but not very heavy: glass/epoxy; (3) inclusion which is very stiff as well as very heavy: steel/Plexiglas and (4) inclusion which is very light and very compliant: glass microballoons/Plexiglas. The frequency was varied over two decades $0.15 \le n \le 10$ MHz. The normalized frequency varied in the range $0.05 \le k_1 a \le 10$. Thus all three regimes of interest were examined: wavelengths large, comparable, and small compared to the inclusion radius. The volume fraction of inclusions was varied in the range $5 \le c \le 50$ percent nominal. The constituent properties are listed in Table 1 below.

TABLE 1—PROPERTIES OF THE CONSTITUENTS						
Material	Longitudinal Velocity c_1 mm/μs	Shear Velocity c_2 mm/μs	Young's Modulus E GPa	Poisson's Ratio ν	Specific Gravity ρ	Inclusion Radius a (mm)
Glass	5.28	3.24	62.77	0.200	2.492	0.15, 0.5
Epoxy	2.54	1.16	4.309	0.3702	1.1180	1.0, 1.5
Lead	2.21	0.86	23.567	0.411	11.3	0.66
Epoxy	2.64	1.20	4.689	0.3715	1.202	
Steel	5.94	3.22			7.8	0.55
Plexiglas	2.63	1.32			1.16	

Fig. 1—A typical toneburst through a specimen

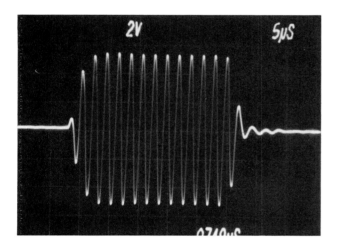

Experimental Procedures

A detailed description may be found in Ref. 1. A very brief description is included here. The heart of the system is a pair of accurately matched, broadband, piezoelectric, water-immersion, ultrasonic transducers; one acts as the transmitter, the other as the receiver. The specimens are 50 X 50 mm square with thickness direction parallel to the direction of wave propagation in water. We emphasize that in the formula $E = \rho c^2$, c is the phase velocity. Therefore, it is critically important to measure the phase velocity and not the group velocity.* In order to ensure that we

---

* Phase velocity, c, is the speed with which a point of constant phase moves through a medium. The group velocity is the speed with which energy flows in the medium. For nondispersive media, e.g. metals, the two are identical and one need not worry about the distinction between the two. However, for dispersive media, such as particulate composites under consideration, the two can be substantially different. In fact the group velocity can be negative!

were measuring the phase velocity, the following precautions were observed. Figure 1 shows a typical toneburst through a composite specimen. Attention was focused on a reference peak near the center of the toneburst (typically the tenth peak from the head of the toneburst). The arrival time of the reference peak at the receiving transducer through water only (t_0) was measured. Now a specimen of thickness w_1 was introduced in the wavepath and the arrival time of the reference peak was measured again (t_1) The same procedure was repeated with a specimen of thickness w_2(time t_2). The time interval between successive peaks in the neighborhood of the reference peak was measured (accuracy = ± 1 nanosecond) for all three cases. It was ensured that the phase varies by 2π between successive peaks. From these measurements the wavespeed can be calculated in three different ways. With V_w as the velocity of sound in water, they are

$$\frac{1}{<c_1>} - \frac{1}{V_w} = \frac{t_2 - t_1}{w_2 - w_1} \qquad (1)$$

$$\frac{1}{<c_1>} - \frac{1}{V_w} = \frac{t_1 - t_0}{w_1} \qquad (2)$$

$$\frac{1}{<c_1>} - \frac{1}{V_w} = \frac{t_2 - t_0}{w_2} \qquad (3)$$

The data was deemed acceptable only when the three calculations agreed.

Finally, and perhaps most importantly, the wavespeed is independent of the choice of the reference peak. Therefore, the wavespeed c_Q (the subscript Q stands for questionable) was plotted as a functon of the peak number (the first little peak is assigned number 1 and so on). As expected, there were some fluctuations in c_Q near the head of the pulse which consists of the transients (here the motion is not purely time harmonic but consists of a range of frequencies). However, within a few cycles c_Q becomes independent of the peak number ensuring that we are, in fact, measuring the phase velocity and not the group velocity.

We reproduce below the formulas by which one can calculate the elastic moduli, E and ν, from the two wavespeeds, c_1 and c_2.

$$E = \mu(3\lambda + 2\mu)/(\lambda + \mu) = \rho c_2^2(3c_1^2 - 4c_2^2)/(c_1^2 - c_2^2)$$

$$\nu = \lambda/\{2(\lambda+\mu)\} = (c_1^2 - 2c_2^2)/\{2(c_1^2 - c_2^2)\}$$

Here, λ and μ are the familiar Lame' constants.

$$c_1^2 = (\lambda + 2\mu)/\rho$$

$$c_2^2 = \mu/\rho$$

98

Results

In Figure 2 we show the normalized phase velocity, $\langle c_1 \rangle / c_1$, for the glass/epoxy composite as a function of volume fraction of inclusions, $\tilde{c}$, at n = 0.4 MHz. *Here $\Omega \equiv k_1 a = 0.15$ i.e. wavelength is about 50 times larger than the inclusion radius.*

Fig. 2—Normalized phase velocity as a function of volume fraction of inclusions for a glass/epoxy composite

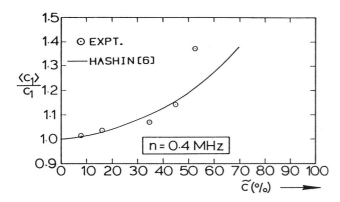

Hashin and Shtrikman[6] have calculated the bounds on the effective elastic moduli of a two-phase (composite) material where the inclusion shape is arbitrary. The solid line in Fig. 2 is the lower bound which is the appropriate bound for the present case where the inclusions are stiffer than the matrix. (When the inclusions are more compliant than the matrix, the upper bound becomes the appropriate bound). The comparison between the (static) theory and the (dynamic) experiment is considered excellent. This figure illustrates the validity of the tacitly made assumption that one can deduce *static* moduli from *dynamic* experiments. We will, however, prove in the sequel that this is true only when the wavelength is large.

Similar results for the lead/epoxy system[2] are shown in Fig. 3. Once again, the solid lines are the upper and lower bounds described by Hashin and Shtrikman. The broken lines are improved bounds described by Miller[7] for *spherical* inclusions; these too are static. Here $k_1 a = 0.15$ and 0.30. Similar results for the case of hollow-glass microballoons in a Plexiglas matrix[3] are shown in Fig. 4; here $k_1 a = 0.11$.

For the convenience of the reader we reproduce here the final results of the calculation by Hashin and Shtrikman.

Fig. 3—Normalized phase velocity as a function of volume fraction of inclusions for a lead/epoxy composite

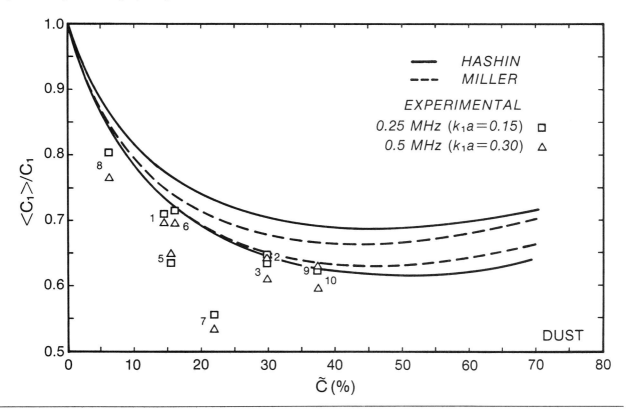

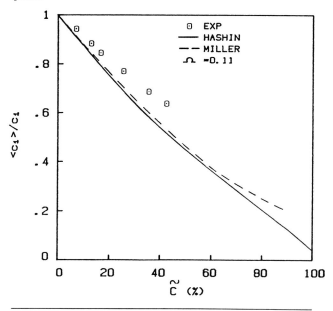

Fig. 4—Normalized phase velocity as a function of volume fraction of inclusions for a hollow glass microballoons/Plexiglas composite

$$K_1^* = K_1 + \cfrac{\mathcal{C}^{\nu}}{\cfrac{1}{K_2-K_1} + \cfrac{3(1-\mathcal{C}^{\nu})}{3K_1+4G_1}}$$

$$K_2^* = K_2 + \cfrac{(1-\mathcal{C}^{\nu})}{\cfrac{1}{K_1-K_2} + \cfrac{3\mathcal{C}^{\nu}}{3K_2+4G_2}}$$

$$G_1^* = G_1 + \cfrac{\mathcal{C}^{\nu}}{\cfrac{1}{G_2-G_1} + \cfrac{6(K_1+2G_1)(1-\mathcal{C}^{\nu})}{5G_1\,(3K_1+4G_1)}}$$

$$G_2^* = G_2 + \cfrac{(1-\mathcal{C}^{\nu})}{\cfrac{1}{G_1-G_2} + \cfrac{6(K_2+2G_2)\mathcal{C}^{\nu}}{5G_2\,(3K_2+4G_2)}}$$

Here, $K_1^*(G_1^*)$ and $K_2^*(G_2^*)$ are upper and lower bounds on the bulk (shear) modulus, respectively; subscripts 1 and 2 refer to the matrix and the inclusions, respectively; $\mathcal{C}^{\nu}$ is the volume function of inclusions; and K and G are given in terms of E and ν by

$$K = E/3\,(1-2\nu)$$

$$G = E/2\,(1+\nu)$$

Next, we illustrate what happens when the wavelength becomes *small* compared to the inclusion radius. Results for the case of glass/epoxy[4] are in-

cluded in Fig. 5; here $k_1a = 3.71$, 4.95 and 7.42, i.e., $k_1a \gg 1$. Note that in going from the long-wavelength case in Figs. 2 and 3, where the data follow the lower bound, to the short-wavelength case here, the data have *exceeded* the upper bound (statically, these bounds cannot be violated). Therefore, at short wavelengths, the modulus calculated from the equations of the type $E = \rho c_1^2$ ceases to have any resemblance to the true *static* modulus. If one ignores this, the error in Young's modulus in Fig. 5 will be of the order of 100 percent. A much more dramatic illustration of the same concept was observed for the case of lead inclusions[2]; this is shown in Figure 6. Note how much the data exceed the upper bound. *Here, the maximum error in modulus would be about 200 percent at $\mathcal{C}^{\nu} = 40$ percent.*

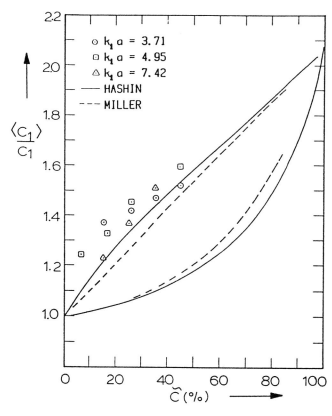

Fig. 5—Normalized phase velocity as a function of volume fraction of inclusions when wavelength small in comparison to inclusion radius, for glass/epoxy composite

In the following we explore explicitly the influence of frequency on the measured wavespeed.

In Fig. 7 we have plotted $<c_1>/c_1$ as a function of frequency *for a fixed-volume fraction*[2]. The arrow STATIC is the lower Hashin bound. As frequency increases, at first the velocity decreases, then it takes a sharp positive jump and becomes frequency indepen-

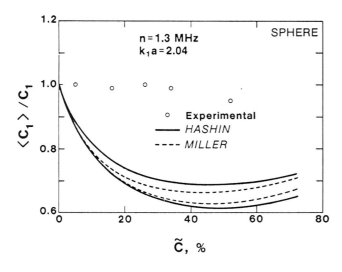

Fig. 6—*Normalized phase velocity as a function of volume fraction of inclusions at high k_1a for lead/epoxy composite*

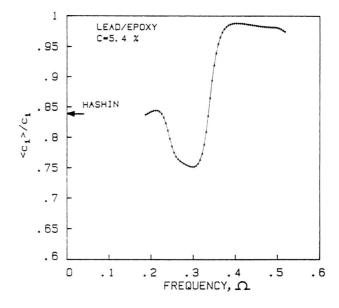

Fig. 7—*Normalized phase velocity as a function of frequency at a volume fraction, $\tilde{c} = 5.4$ percent, for lead/epoxy composite*

dent at higher frequencies. This figure helps us make the main point of this paper: in order to make valid modulus measurements using ultrasound, the normalized frequency, k_1a, must be small compared to one, or the wavelength must be large compared to the characteristic length(s) of the composite. Similar results were obtained for the glass/epoxy system[4] and are shown in Figure 8.

Finally, we made the problem more interesting (and considerably more complicated) by introducing a second characteristic length in the problem. We arranged steel spheres in a *periodic array* in a Plexiglas

Fig. 8—*Normalized phase velocity as a function of frequency at a volume fraction, $\tilde{c} = 15$ percent, for glass/epoxy composite*

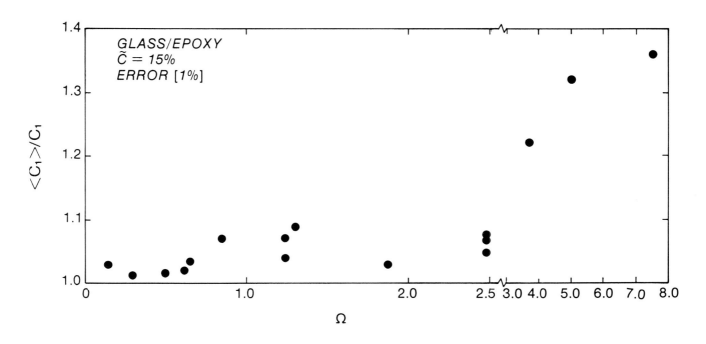

Fig. 9—Steel spheres in a periodic array in a Plexiglas matrix, top and side views.

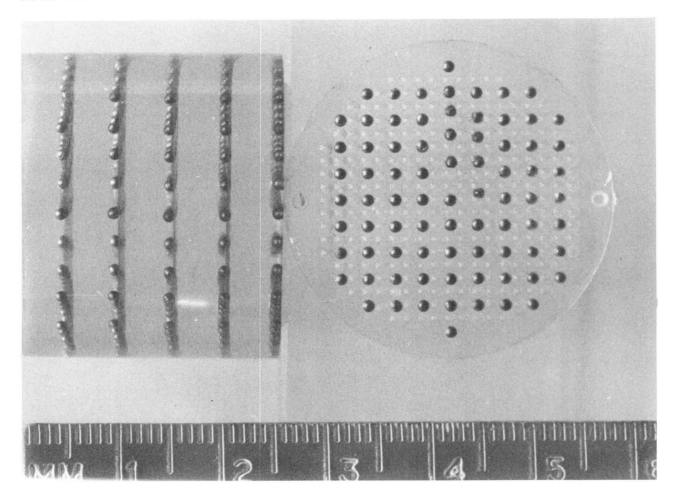

matrix[5]; see Fig. 9. In addition to the radius of the sphere, a, the second characteristic length is the particle spacing in the direction of wave propagation, d. Furthermore, what makes this problem even more interesting is the fact that the two characteristic lengths are of the same order of magnitude: $a = 0.55$ mm, $d = 2.63$ mm, and $d/2a \sim 2.5$. Attention is first drawn to the filled circles (periodic case) in Fig. 10. We note that in keeping with the solid-state physics literature, here the frequency is normalized with respect to d (and not to a): $\Omega = k_1 d/\pi$. As $\Omega \to O$ the data approaches the static value, as expected. However, as Ω increases, the wavespeed changes dramatically. It first decreases (the first pass band), then jumps across the first stop band, again decreases in the second pass band, jumps across the second stop band, and decreases again in the third pass band. Clearly only the low-frequency limit of the phase velocity corresponds to the static modulus. To guard against

fortuitous results, we also manufactured specimens which were exactly identical in all other respects except that the spheres were distributed in a *random* manner; the results are shown as open circles in Fig. 10. It is extremely interesting to note that when Ω is small, i.e., wavelength is large compared to both characteristic lengths of the problem, then the two sets of data converge. The same is true at the other extreme, when $\lambda \ll a$ as well as d. In between, when $\lambda \sim a$ and d, there are significant differences in the dispersion curves. Here we learn yet another interesting lesson. When the wavelength is large, the wave does not 'see' the details of the microstructure, it does not distinguish between the periodic and the random cases (all other things being equal). When the wavelength becomes very small, it begins to 'see' individual sphere and, once again, does not see the difference between the random and the periodic case.

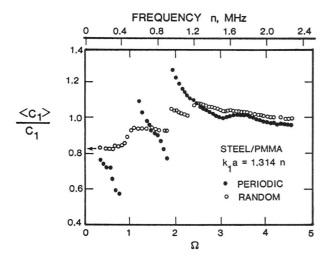

Fig. 10—Pass bands and stop bands in periodic and random steel/PMMA composites

posite material, a. Calculate $\Omega = 2\pi na/c_1$ for the particular test frequency. If $\Omega \leqslant 0.1$, proceed with confidence; the dynamic measurement, ρc^2, is an accurate estimate of the static modulus E. If $0.1 \leqslant \Omega \leqslant 1.0$, one may still be able to obtain valid results; read Refs. 1-5 for details. If $\Omega > 1$, the results are most likely invalid.

Conclusion

With the viewpoint of field applications in mind, we conclude with a rule of thumb for the practitioner. Estimate the characteristic length of the com-

References

1. Kinra, V.K., Petraitis, M.S., and Datta, S.K., "Ultrasonic Wave Propagation in a Random Particulate Composite," Int. J. Solids Struct., 16, 301-312 (1980).

2. Kinra, V.K., "Acoustical and Optical Branches of Wave Propagation in an Epoxy Matrix Containing a Random Distribution of Lead Inclusions," Review of Progress in Quantitative Nondestructive Evaluation, ed. D. O. Thompson and D.E. Chimenti, Plenum Publishing Corp., NY, Vol. 3B, 983-991 (1984).

3. Kinra, V.K., and Ker, E.L., "Effective Elastic Moduli of a Thin-Wall Glass Microsphere/PMMA Composite," J. Comp. Mat., 16, 117-138 (March 1982).

4. Kinra, V.K., and Anand, A., "Wave Propagation in a Random Particulate Composite at Long and Short Wavelength," Int. J. Sol. Struct., 18 (5), 367-380 (1983).

5. Kinra, V.K., and Ker, E.L., "An Experimental Investigation of Pass Bands and Stop Bands in Two Periodic Particulate Composites," Int. J. Sol. Struct., 19, 393-410 (1983).

6. Hashin, Z. and Shtrikman, S., "A Variational Approach to the Theory of the Elastic Behavior of Multiphase Materials," J. Mech. and Phys. of Sol., 11, 127-140 (1963).

7. Miller, M.N., "Bounds for Effective Bulk Modulus of Heterogeneous Materials," J. Math. Phys., 10, 2005 (1969).

Section VB

Acoustic Emission in Composites

by R.A. Kline

Introduction

The term 'acoustic emission' (AE) is applied in a broad sense to the sounds which are internally generated in a material which is under stress. AE testing is relatively simple with only a sensor (usually piezoelectric), basic signal-analysis equipment (amplifier, filter, data-processing equipment) and some means of loading the structure needed. There is a wide variety of mechanisms known to be responsible for AE generation, ranging from dislocation motion to crack propagation. Therefore, the basic AE monitoring techniques are applicable to a wide class of materials and structures. This is particularly true for composites with several mechanisms, including fiber fracture, fiber-matrix debonding, matrix crazing and cracking, and delamination known to generate AE signals in composites.

One of the principal advantages of AE inspection lies in its ability to effectively cover a wide area in a relatively short amount of time as compared with other NDT methods. This is attributable to the fact that in most structures, sound-wave propagation is quite efficient, with sound waves traversing several meters without experiencing undue attenuation. Unlike ultrasonics, where a single transducer must be mechanically scanned over the structure, once installed, a fixed array of transducers can be used to rapidly inspect the same structure and, with suitable precautions, be left in place to monitor the performance of the structure in service over long periods of time. This capability makes AE attractive for the inspection of large composite structures like pressure vessels, aircraft wing sections, etc. While most composite applicatons have involved polymer-matrix composites[1], there is a growing body of research and development work with metal-matrix composites[1-2]. The basic methods of AE testing are equally applicable to both composite media.

Naturally, there are limitations to the applicability of AE techniques, some of which are very important. With AE, spatial-flaw location is limited to a general location within the structure. This is in contrast to ultrasnonics or radiography, where a detailed two-dimensional projection or shadow image of the flaw is presented. Nor is there any way to assess the severity of a defect based solely on AE criteria. A large amount of AE activity may come from a relatively benign flaw with less activity observed from a more serious flaw. Precise source identification, while theoretically plausible, is as yet highly impractical in all but the most carefully controlled laboratory settings. Composite materials present special problems to the AE investigator. Specimen anisotropy makes source location more difficult in composites than it is in isotropic materials. Signal attenuation is often higher in composites than in other structural materials, necessitating the use of larger number of sensors in the inspection. Despite these drawbacks, AE test methods remain exceptionally useful for composites. This is particularly true for large scale structures where AE can quickly identify suspected problem areas for more detailed inspection with an alternative technique.

Signal Analysis

Acoustic-emission signals generally occur as individual transient signals known as burst emission (Fig. 1). These signals are usually characterized by a rapid rise to a maximum followed by an exponentially decaying, osciallatory response (Fig. 1a). For some processes (e.g., yielding in metals), the emissions appear to form a single signal. This is known as continuous emission. In all likelihood, continuous emissions represents several discrete events, closely spaced in time, so that preceding signals do not have sufficient time to decay completely before subsequent events occur. Continuous emission signals are illustrated in Fig. 1(b).

A variety of techniques have evolved to quantify the characteristics of AE activity. These methods are illustrated in Fig. 2(b) for a prototype AE signal. One of the simplest and most widely utilized methods of signal parameterization remains AE counts. Signal counts represent the number of excursions of a signal above a fixed threshold level. It should be pointed out that AE counts, like all of the readily available signal-characterization schemes, are not readily interpreted in terms of more familiar measures. The number of counts associated with a given AE event will be a function of not only the strength of the AE source signal, but the sensitivity of the transducer, its coupling efficiency, electronic signal amplification,

Fig. 1 (a)—Simulated burst emission,

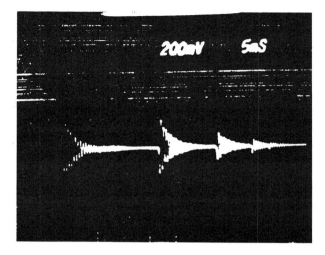

(b) Simulated continuous emission

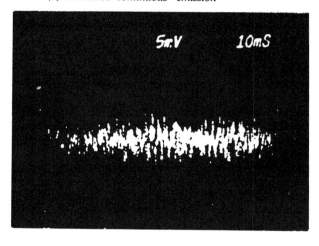

signal location, specimen attenuation and threshold-level setting. Therefore, one should exercise caution in interpreting the significance of this (or any other quantitative AE measure) except in closely controlled circumstances.

Other common signal-characterization parameters are also illustrated in Fig. 2. They include: *amplitude— usually the voltage level of the initial excitation A_o; energy*—a measure of the total energy present in the total signal $\propto A^2$. May be analog (RMS) or digital; *rise time*—time required for signal to reach initial maximum τ; *duration*—time required for signal to decay to a fixed threshold level; and *event*—the number of AE events (independent of amplitude or duration) above a fixed threshold value. In many cases, these measurements are shown graphically in one of two ways: totaled (as in total counts for the complete test to any given point in time) or as rates (as in counts/unit time interval). Naturally, these parameters are all interrelated. For

example, assuming an exponentially decaying sinusoidal oscillation as the typical AE signal of constant frequency ω:

$$(1) \qquad V(t) = V_o\, e^{-\gamma t} \sin \omega t$$

where

$V(t)$ = output voltage of sensor

V_o = initial signal amplitude

γ = decay constant (> 0)

ω = signal frequency

The number of counts above a fixed threshold V^* will be given by

$$N = \frac{t^*}{2n/\omega} = \frac{\omega}{2\pi\gamma}\, ln\, \frac{V_o}{V^*} \qquad (2)$$

where

$$V^* = V_o e^{-\gamma t^*}$$

and

$$5^* = \frac{1}{\gamma}\, \propto n\, \frac{V_o}{V^*}$$

As mentioned earlier, AE signal parameters are influenced to experimental conditions as well as the actual AE signals. As such, it is difficult to estimate flaw severity on AE results alone.

AE Instrumentation

Sensor

Sensors for ultrasonic applications have already been discussed in this volume. Since AE transducer requirements are somewhat similar, many of the same considerations apply. However, there are several important differences which need to be discussed. For AE applications, one is principally concerned with transducer sensitivity (hence resonant rather than wide-band sensors), particularly in the frequency region from 100 kHz-500 kHz where the largest fraction of AE activity is observed. It should be noted that, while higher frequency components are probably present in AE signals, internal signal attenuation usually precludes these higher frequency components from being observed. This is in contrast to the 1-mHz+ frequencies required for adequate flaw definition in ultrasonic testing. While ultrasonic transducers are scanned, AE transducers will be affixed to test articles, requiring permanent mounting. Coupling considerations are much the same as those for contact ultrasonics.

Fig. 2—Acoustic-emission signal parameters

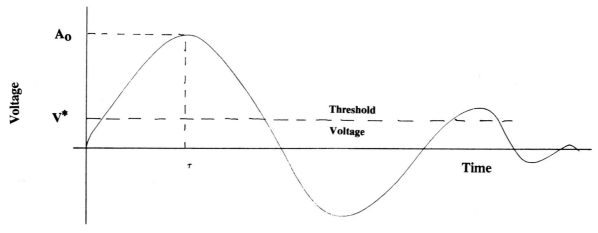

RING DOWN COUNTS - Excursions above threshold voltage, usually displayed as either total counts or count rate.
EVENT COUNTS
SIGNAL AMPLITUDE - A$_o$
SIGNAL RISE TIME - τ
ENERGY CONTENT - Two approaches: Analog (R.M.S., Direct)
           Digital

A variety of special-applicaton transducers are also available. For frequency analysis, it is desirable to use heavily damped sensors with their increased bandwidth, despite the attendant sacrifice in sensitivity. Sensors are also commercially available for use in extreme environments (e.g., high temperature) which use sensing elements with a higher Curie point than conventional materials (e.g., lithium niobate) and temperature-resistant backing materials. A novel sensor has also been developed for making measurements of the normal component of the surface displacement at a point[4]. This is in contrast to conventional piezoelectric plate sensors which produce an average response over the entire transducer surface. This type of transducer is illustrated in Fig. 3 and represents a relatively accessible way in which surface displacement can be measured without resorting to more complicated devices like the capacitive sensor or opitcal interferometer. The principal application of this device is in AE source parameter measurement.

Electronics

A schematic of the basic electronic system used in typical AE testing is shown in Fig. 4. As illustrated here, the electronics needed are relatively limited,

Fig. 3—NBS point displacement transducer (after Proctor, Ref. 4)

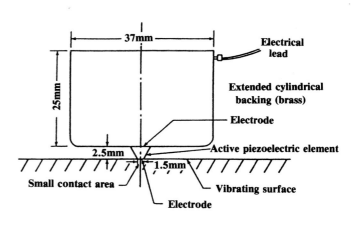

with only a preamplifier (typical gain of 20-40dB), amplifier (for an additional 40-60 dB gain) signal-measurement device (e.g., counter), and recording instrument being required. Amplifier bandwidths usually range from 50 kHz to ~ 1mHz. A bandpass filter is optional but often useful to reduce signal

Fig. 4—Electronic system diagram

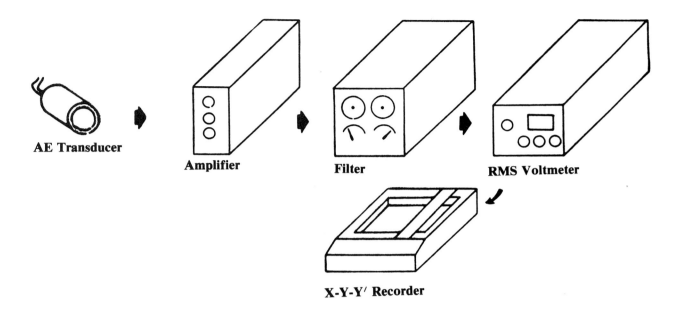

AE Transducer   **Amplifier**   **Filter**   **RMS Voltmeter**

X-Y-Y′ Recorder

noise from extraneous sources such as specimen vibrations, part rubbing, etc. which are usually lower in frequency than typical AE events.

Signal¢analysis equipment ranges from simple to sophisticated. For many AE applications, only a counter with a variable threshold is needed. An RMS meter is also a useful way to characterize AE activity, particularly when 'continuous' AE activity is observed. Commercial AE equipment usually employs a microprocessor chip which allows the user to select virtually any signal parameter at will for display. Many of these devices allow detailed signal characteristics to be recorded in memory for each AE event. Post-test processing yields a display of the time evolution of any AE parameter desired. Several of these devices employ multiple sensors/channels which allows for the location of the Ae source to be identified (see source-location section). An additional advantage of multiple sensors is that signals from non-AE sources (as indicated by their relative arrival times) may be excluded from the data analysis. Strip-chart recorders are often useful for recording data from long-term AE tests.

AE Source Location

The utility of AE testing for large-scale structures relies upon its source-location capability. This can be best illustrated by the one-dimensional source-location example shown in Fig. 5. An AE signal is shown to originate at some unknown position X

(measured from transducer 1) within the sample at time t_o. The signal will then arrive at the two transducers at times $t_o + t_1$ and $t_o + t_2$, respectively. If the velocity of propagation in this one-dimensional example is known (say V_o), then X can be determined by simple algebra as follows:

$$X = V_o 5_1$$
$$L - X = V_o t_2 \tag{3}$$
$$X = \frac{L - V_o \Delta t}{2}$$

where $\delta t = t_2 - t_1$.

For a two-dimensional plate structure, the situation is somewhat more complex but tractable for isotropic media using triangulaton techniques as illustrated in Fig. 6. If one uses a three-transducer array and works through the algebra, it is found that the locus of possible sources with a given time delay between any pair of transducers forms a hyperbola in space. Thus, by measuring the time delays between two sets of transducers and constructing the associated hyperbolae, the AE source is found as the intersection of the two curves.

For anisotropic media, the problem is significantly more complicated. Since the material properties are directionally dependent, propagation velocities will vary with propagation direction. This means that the

108

Fig. 5—One-dimensional AE source location

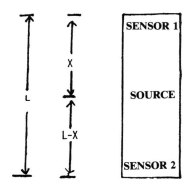

Fig. 5—One-dimensional AE source location

Fig. 6—Two-dimensional AE source location

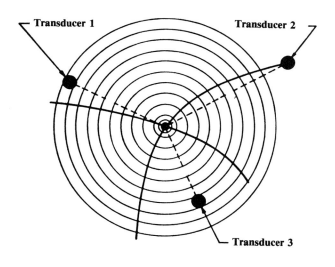

curves representing possible source locations will be distorted from their hyperbolic shape in isotropic media. While source triangulation is still feasible, the increased mathematical complexity makes such calculations impractical for most test purposes. Alternatively, an approximate source-location method called the zone system is more commonly utilized for composite structures. This method, illustrated in Fig. 7, uses relative signal amplitudes and known signal attenuation characteristics to yield an approximate location of the source. While not as accurate as triangulation, the method has proved quite useful for composite testing.

Fig. 7—Zone location

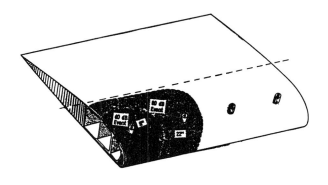

The Kaiser and Felicity Effects

In early AE tests of many samples, it was observed that acoustic-emission activity, which was present during the initial sample loading, was not observed upon subsequent reloadings of the sample until the previous maximum applied load was reached. This phenomena was originally observed by J. Kaiser[5] in the early 1950s and bears his name today as the Kaiser effect (Fig. 8). This phenoma has been utiliziea for several structures as an estimate of the maximum load that the structure has experienced in its service environment. This is achieved by instrumenting and loading proof the structure in a known fashion to the point where AE activity is initially observed.

Subsequent research has demonstrated that the Kaiser effect is not a universally observed phenomena. Unless the sample is metallic, unflawed, and the reloading is immediate, the Kaiser effect may not be observed. This is particularly true for composite materials. For composites, because of the heterogeneous composition, failure surfaces will be jagged and AE from sources like rubbing of crack surfaces is likely to occur during retesting. The lack of the Kaiser effect is not entirely disadvantageous. Fowler[6,7] and others have exploited this as a means

Fig. 8—Kaiser effect

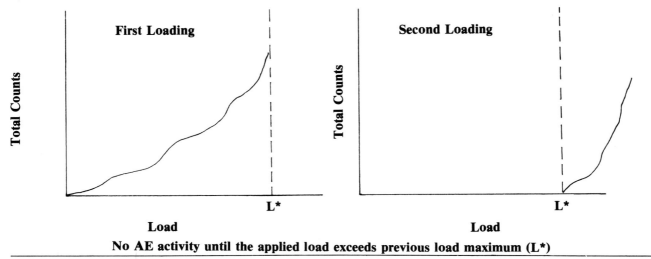

No AE activity until the applied load exceeds previous load maximum (L*)

of estimating damage in composite materials. This effect is quantified as the Felicity ratio (Fig. 9), defined as follows:

Felicity ratio =

$$\frac{\text{load where AE is 1st observed on reloading}}{\text{previously applied maximum load}} < 1$$

This measure has been found to provide a means of monitoring damage development[8] in fiber-reinforced composites. Felicity ratio measurements are also commonly used to monitor the performance of filament-wound pressure vessels[9].

AE Source Discrimination

One of the potential advantages of AE is the ability to discriminate between various possible generation mechanisms. This capability is particularly important for composite materials where acoustic emission may arise from several sources including fiber fracture, fiber-matrix debonding, matrix cracking, and ply delamination. Since the severity of those flaws are

Fig. 9—Felicity effect

(Analog to Kaiser Effect for Composite Materials)

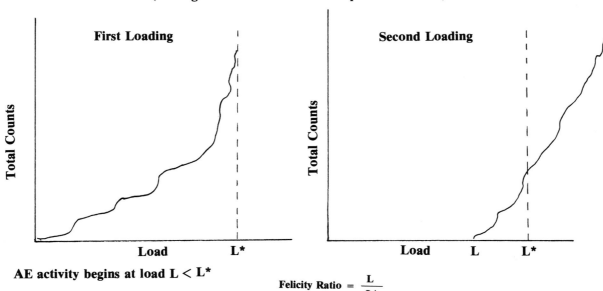

AE activity begins at load L < L*

$$\text{Felicity Ratio} = \frac{L}{L^*}$$

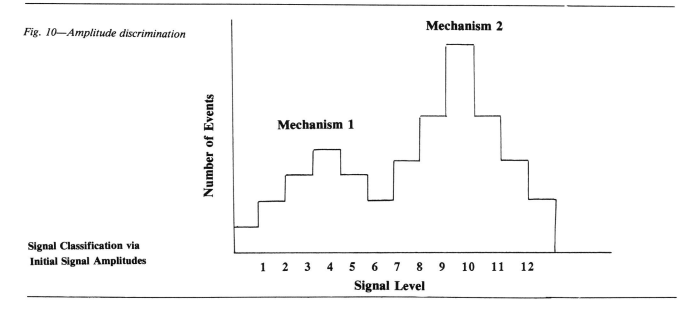

Fig. 10—Amplitude discrimination

Mechanism 2

Mechanism 1

Number of Events

Signal Classification via Initial Signal Amplitudes

1 2 3 4 5 6 7 8 9 10 11 12

Signal Level

likely to be different, the value of source discrimination is obvious. However, efforts to classify the sources of AE signals based on observed signal features has, thus far, met with limited success. The reaons for this are manifold, ranging from the complexity of the signal-generation process, the effects of the propagation medium on the signals and sensing limitations. Several methods have been suggested and utilized for AE source discrimination. They include: *amplitude-distribution analysis*—signal classification based on histogram plots of measured signal amplitudes (Fig. 10); *frequency analysis*-signal classification based on spectral content of AE signals (Fig. 11); and *source parameter measurement*—signal classification based on detailed analysis of AE waveform to extract data about the actual AE source (Fig. 12). Of these, only amplitude distribution has progressed beyond the laboratory curiosity

stage and is described below. For more information regarding alternative source characterization techniques, the reader is referred to Ref. 10.

Amplitude-distribution analysis was the first source-discrimination technique proposed and remains the most widely utilized approach. Amplitude discrimination is based on the expectation that AE signal strengths will depend on the source mechanisms. If one measures the amplitude from each AE event and places it in one of several predetermined categories based on its amplitude, histogram plots like that shown in Fig. 10 may be developed. It is then hoped that a multimodal amplitude distribution will emerge (bimodal in this illustration) with the signals in each mode corresponding to a different mechanism. Wadin has utilized this approach to examine AE amplitudes from three-

Fig. 11—Frequency analysis

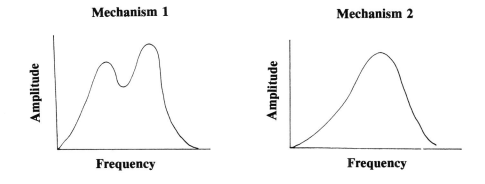

Mechanism 1

Mechanism 2

Amplitude

Amplitude

Frequency

Frequency

Signal Characterization via Spectral Signature

111

Fig. 12—Source-parameter measurement

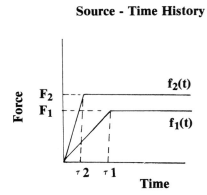

Source - Time History

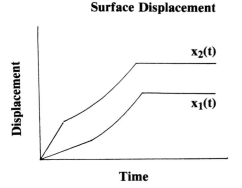

Surface Displacement

Quantitative Source Characterization via Measurement of τ & F From Surface Displacement Measurements

point bend tests in composites[11]. As shown in Fig. 13, a trimodal distribution was found which the authors ascribed to matrix cracking, fiber-matrix debonding and fiber fracture. Efforts are underway to substantiate these claims. The principal advantage of the amplitude-discrimination approach is its simplicity,

both in terms of equipment and data analysis. However, it should be mentioned that this is an imperfect technique. Amplitude results will be highly sensitive to variations in the sensor placement, coupling efficiency, AE spectral content and sensor resonances, etc.

Fig. 13—Amplitude-distribution (after Wadin, Ref. 11)

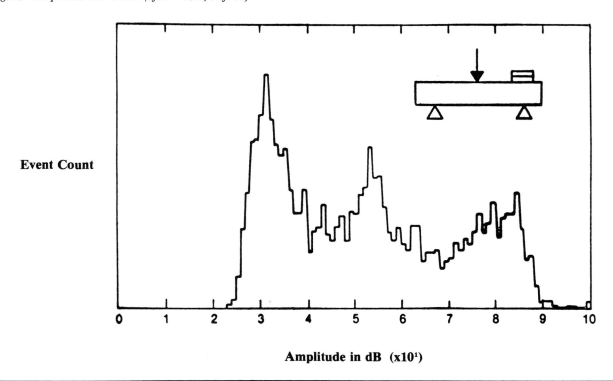

Acousto-Ultrasonics

Acousto-ultrasonics, pioneered by A. Vary of NASA Lewis[12], represents a hybrid approach to composite NDE, combining some of the features of acoustic emission and ultrasonic test methods. With this approach, a pair of ultrasonic transducers are used to generate and sense elastic waves (dominated by plate modes in laminated composite plates) in the composite structure. Hence, the technique is somewhat akin to the ultrasonic testing approach where externally, rather than internally, generated waves are used; however, guided waves rather than bulk waves are now the dominant components of the motion. Acoustic-emission signal-characterization techniques (counts, amplitude, energy, etc.) are then used to quantify the signals. Since guided waves are used, the approach yields a measure of the attenuation present in the sample along the path between transducers. With suitable precautions, measurement reproducibility can be insured. Henneke and coworkers at VPI have recently developed a system which can be used to scan composite structures with this approach[13]. Results, to date, with acoustoelastic testing are quite promising.

References

1. Hamstad, M., "Testing Fiber Composites with Acoustic Emission Monitoring," J. Acoust. Emis., 1, 151-164 (1982).
2. Johnson, C., Ono, K., and Chellman, D., "Study of Mechanical Behavior of Metal-Matrix Composites Using Acoustic Emission," Jap. Soc. Comp. Mat., 647-655 (1986).
3. Johnson, C., Ono, K., and Chellman, D. "Acoustic Emission Behavior of Metal-Matrix Composites," J. Acoust. Emis., 4, S263-269 (1985).
4. Proctor, T., "Some Details of the NBS Conical Transducer," J. Acoust. Emis., 1, 173-178 (1982).
5. Kaiser, J., PhD thesis, Techn. Hoc., Munich (1950).
6. Fowler, T., "Development of an Acoustic Emission Test for FRP Equipment," ASCE Ann. Conv., Boston, MA, Preprint 3583 (1979).
7. Couslik, D., and Fowler, T., Composite in Pressure Vessels and Piping, ed. S.V. Kulkarni and C.H. Zweben, ASME, New York, 1-16 (1977).
8. Kline, R., "Effect of Microstructure on the Mecanical Behavior of Sheet Molding Compound Composites," Composite Materials: Quality Assurance and Processing, ed. C. Browning, ASTM, 133-156 (1983).
9. Golaski, L., Rumosa, M., and Hull, D., "Acoustic Emission Testing of Filament Wound Pipes Under Repeated Loading," J. Acoust. Emis., 1, 95-101 (1982).
10. Wadley, H.N.G., and Scruby, G., "Acoustic Emission Source Characterization," Advances in Acoustic Emission, ed. H. Dunegan and W. Hartman, Dan Hunt Publishing, Knoxville, 125-154 (1981).
11. Wadin, J., Acoustic Emission Applications, Dunegan-Endevco, San Juan Capistrano, CA (1978).
12. Vary, A., "Acousto-Ultrasonic Characterization of Fiber Reinforced Composites," Mat. Eval., 40, 650-654 (1982).
13. Henneke, E., Sundaresan, M., and Debuct, M., "NDE of Composite Spherical Pressure Vessels by Acousto-Ultrasonics," ASNT Spring Conf., Phoenix, AZ (1986).

Section VI A

Detection of Damage in Composite Materials Using Radiography

by R. Van Daele, I. Verpoest, and P. De Meester

Introduction

Many different material properties must be examined when using composite materials due to their heterogeneous nature. Structural applications such as use of laminates, tubes and vessels require meticulous quality inspection. Important factors include fiber concentration, fiber orientation, resin porosity and section thickness. In addition to general material properties, detailed information on the microscopic level is sometimes desired, i.e., the presence of matrix cracks, delaminations, fiber breakage and debonding. X-ray radiography can reveal most of these properties and defects and has been used successfully by numerous researchers.[1,2] The other damage types such as debonding and fiber fracture can be detected under specific conditions.

Many general publications and detailed information have been written on the production of X rays and X-ray equipment, and will not be repeated here.[3] Discussion will be limited to the information necessary to produce good radiographs of composite materials with a polymer matrix. The typical X-ray conditions for composite materials, the specimen preparation (penetration) and information on sensitivity will be given. However, this document will not take into consideration special benefits and limitations resulting from the use of nonfilm recording media such as paper, fluoroscopy, xeroradiography, and electronic image recording and processing. Readers unfamiliar with nomenclature can find these explained in the appendix.

The first major section of this chapter will deal with low-voltage radiography without the use of X-ray opaque penetrants, showing the conditions and possibilities of this technique. The second section will indicate the advantages of the use of penetrants and will discuss the choice of the penetrant and penetration conditions. Typical X-ray conditions will be given and comments on the quality of the obtained radiographs will be made. Finally, special techniques like microradiography and steroradiography will be discussed.

Low-Voltage X-Ray Radiography (Without Penetrant)

Producing a Suitable X-ray Spectrum: Low Voltage to Increase Contrast

Defects in composite materials (1-5 mm thick) are very small. Also, the material contains only elements with low atomic number and density and thus low X-ray absorption. Therefore, it is necessary to reduce the energy of the radiation in order to maintain object contrast. High X-ray energy results in lowered

Fig. 1—Density values at different voltages: (+)10 kV 150 min; (*) 14 kV 5 min, (x) 16 kV 120 s ffd = 50 cm, 10 mA, D4

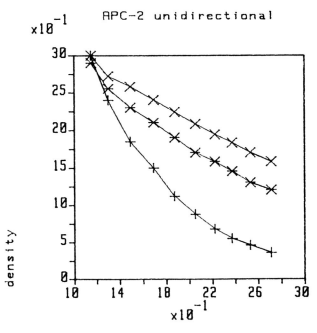

115

differential absorption and increased scatter. Hage-maier measured contrasts 15 percent higher when using 20 kV than when using 40 kV.[10] For thin composites (1-3 mm), voltages in the range of 15-25 kV are used most. As an example, Fig. 1 shows the density obtained at a given exposure for three different voltages. It is clear that at lower voltages (10 kV), the density difference (contrast) is higher than the one obtained at higher voltage (16 kV). Since we are interested in the highest possible contrast, we will use the lowest voltage possible. However, from the other X-ray conditions (exposure time) it is also clear that the 150 minutes of exposure time is far from practical. Therefore a compromise will have to be made between the desired contrast (maximal) and the exposure time (minimal). Figure 2 shows that at a given voltage, the higher the density of the radiograph, the higher the contrast obtained.

Fig. 2—Density values at different exposure times: (+) 40 s; (*) 60 s; (x) 120 s 16 kV, ffd = 50 cm, 10 mA, D4

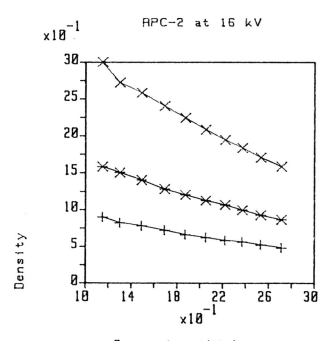

Since the voltage used for composite materials is below 60 kV, several factors must be considered carefully, such as the material between source and object, the material of the cassettes in which the film is placed, and the film and subject contrast.[4] How to make high-quality radiographs and determine the resolution will be discussed below.

Resolution and Detection Limits

In the classic contact microradiography, with the object directly above the film, the individual fibers are below the limit of resolution and the composite appears more or less homogeneous. However, the general texture and mode of lay up are usually clearly observed because the fibers and matrix absorb the X-rays differently (Fig. 3). Particulate inclusions (e.g., parts of the packing paper of prepreg sheet) and matrix-rich zones (Fig. 4) can also be revealed. In addition to overall material distribution, one can easily study fiber-bundle orientation, quality of weave (Fig. 5) and fiber distribution (Fig. 6) using low-voltage radiography. A procedure often used is to include trace fibers of glass or boron in the prepreg in order to discern easily the lay up of the laminate.[5]

The fiber content per unit surface area can be determined easily for glass fibers, after proper calibration due to the high absorption coefficient of glass (about 20 times that of most resins), by measuring the film density.[6] When radiographing carbon fiber-reinforced composites, however, the X-ray absorption of polymers and carbon fibers is very similar (both have a high carbon content), and differs only slightly from that of air (filling the cracks and pores). Martin has shown that it is practically impossible to use X-ray radiography to determine the carbon-fiber content.[20]

Large cracks in the material may be detected using low-voltage radiography, taking into consideration the same precautions on orientation of cracks as in metallic materials: e.g., delaminations are difficult to detect because they are oriented normal to the radiation.

Obtaining the Right Exposure

The density of the exposed radiograph is dependent on a number of factors: material and material thickness, intensity of the X-ray beam, distance between source and object, sensitivity of the film, use of intensifying screens and film processing. How these are controlled and how they affect the quality of the radiograph will be discussed below.

Film

The finest image detail that can be recorded on a film will depend partly on the size and distribution of the developed silver grains in the emulsion. In general, the smaller the grains, the finer the details that can be recorded. The choice of the grain size depends mostly on the amount of detail one is interested in and the amount of exposure time one is willing to spend: the smaller the grains, the longer the exposure. A number of different films with different 'speeds' are commercially available. The value of the film speed has been chosen arbitrarily as '100' for D 4 (Agfa Gevaert). The speed is related to the radiation exposure, measured in rontgen, for a film density of 2.0.

Fig. 3—Radiograph of an eight layer crossply (0, 90)₂ₛ glass-epoxy laminate (Fiberdux 913 G−E−5−30) showing clearly the two directions of the glass fibers (the white lines) embedded in the matrix

Fig. 3—Radiograph of an eight layer crossply $(0, 90)_{2s}$ glass-epoxy laminate (Fiberdux 913 $G-E-5-30$) showing clearly the two directions of the glass fibers (the white lines) embedded in the matrix

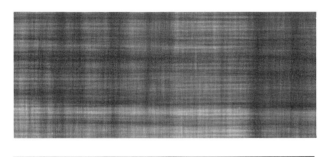

Fig. 4—Radiograph of a $(0_2 90_2)_s$ carbon-epoxy (Fiberdux-920-C-TS-542) showing matrix-rich zones (the more white regions)

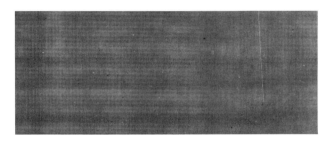

Fig. 5—Radiograph of a woven glass fabric laminate. The arrows indicate zones of bad weaving

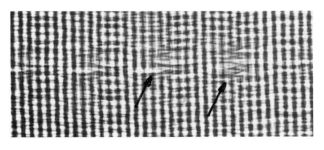

Fig. 6—Radiograph of a filament wound unidirectional glass. The winding stretegy was chosen incorrectly, resulting in inhomogeneous distribution of the glass fibers (white lines)

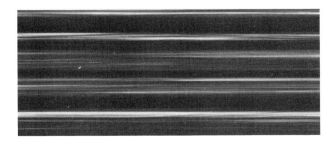

TABLE 1—SPEEDS OF DIRECT TYPE INDUSTRIAL RADIOGRAPHIC FILMS

Film	Speed	Remarks	Film	Speed
KODAK (GB)			*Agfa Gevaert*	
NS-2T	1300		D 10	1000-1200
Industrex D	700-800		D 7	50-400
Industrex CX	300-450		D 5	190
Industrex MX	80		D 4	100
Industrex MX	90	5-min development	D 2	30-50
Industrex MX	100	6-min development		
Industrex MX	130	Automatic processing		
KODAK (USA)			*Dupont-Cronex*	
AA	300		NDT 75	400-500
T	150		NDT 65	250
M	60-80		NDT 55	125
R	30			

X-ray conditions: 200 kV X rays
12-mm Cu filter close to the X-ray tube
lead intensifying screens (0.05 mm front and 0.2 mm back) (from Halmshaw[3])

For composite materials, a film with a speed of about 100 is usually sufficient: it has a good resolution, and the exposure time is acceptable.

Screens

It is common practice in industrial radiography to use intensifying lead (Pb) screens. These screens have two functions: absorbing scattered X-rays and decreasing the exposure time by producing free electrons (by the action of the X-rays) which interact with the emulsion. Using a low voltage of about 20 kV, these Pb foils do not emit free electrons because the X-ray energy is too low. Besides that, they absorb too much radiation and thus increase the exposure time by a large factor. Experiments have shown that at low voltages, a thin plastic sheet is sufficient to absorb scattered radiation; thus the use of plastic envelopes to contain the radiographs is sufficient to remove scattered radiation. A Pb screen behind the film can be used to absorb backscattered radiation; however, experiments by Fassbender and Hagemaier have shown that taking radiographs without a Pb screen behind the radiograph does not influence the quality much, since at these low voltages, the amount of (back) scattered radiation is low and largely absorbed by the plastic container.

The authors found that, when using a voltage of about 10 kV, the use of 'Vacupac, D4' by Agfa Gevaert gave good results if the Pb foil of the exposure side of the film was removed. The plastic envelope did not produce any texture on the X-ray film, while the use of paper envelopes did. The use of fluoremetallic intensifying screens has been investigated by Fassbender and Hagemaier, who observed an exposure-time reduction of 10 percent (10 kV) to 33 percent (20 kV). Whether or not to use these fluoremetallic screens will be determined mainly by the amount of time one wants to spend in exposing the specimen to X-rays and by the quality of the radiograph that is required (see the section, Radiographic Definition).

Film to Focus Distance (ffd)

In medium- (150-400 kV) to high-voltage (400-1000 kV) X-ray applications, the intensity of the X-ray beam decreases inversely with the square of the distance, i.e.:

$$\frac{I_1}{I_2} = \frac{D_2^2}{D_1^2} \qquad (2)$$

where I_1 and I_2 are intensities at the distances D_1 and D_2, respectively. As was mentioned earlier, this is not the case for low voltages (10-60 kV): the lower the voltage, the more the X-rays will be absorbed by the air. Therefore, any changes of the ffd will have an important influence on the exposure time which, unfortunately, cannot be calculated from eq (2). An experimental determination of the exposure time will be necessary. This influence will drastically change when substituting air (X-ray transmission coefficient of 13 percent) by helium (99.9 percent transmission). Helium transmits X-rays more readily than does air. Therefore, in the case of very thin specimens where very low voltages (eg., 7 kV) and large ffd's (eg. 1 m) are desired, placing a balloon or canister (with PET windows) filled with helium between the X-ray source and the specimen can greatly reduce the exposure time required.

Exposure

The radiograph should have a film density between 2.0 and 3.0 or higher if suitable film-viewing equipment is available. This density is obtained by exposing the specimen and film to X-rays, during a given time with a given filament current.

As mentioned above, the inverse-square law for the decrease of the X-ray intensity with distance does not apply for low voltages in air. Thus, calculating the exposure from earlier data becomes rather difficult. Also, there are as yet no exposure charts available to estimate the exposure under given circumstances (e.g., voltage, thickness, film, material, and ffd). Therefore, one must determine the exposure experimentally.

There are three methods which can be used to determine the exposure. (1) The so called 'fixed exposure' method, consists of keeping the film-to-film focus distance (ffd), exposure time and filament current constant while changing the voltage until the desired film density is obtained. This method, however, does not yield optimum contrast sensitivity. (2) The 'optimum energy' method, as described by Wysnewski, consists of determining the 'equilibrium half-value' (EHVL) by measuring the transmission of

the X-rays (in percent) for different voltages and different material thicknesses.[7] Once a graph such as that shown in Fig. 7 has been measured, EHVL values can be determined. For the voltage used, the EHVL is one fifth of the composite thickness (Fig. 8). Adequate exposure times are then experimentally determined. The disadvantage of this method is that one must measure the X-ray dose that is transmitted through the material. This measurement requires extra equipment and time to determine the EHVL values. The big advantage, however, is that once these curves have been made, the voltage values can be determined easily for different material thicknesses, simply by reading the values from the table. (3) Another approach is to use a certain voltage (16 kV for 1-mm thick carbon-epoxy composites has been found to give excellent results) and the maximum filament current and film-to-focus distance as is reasonably possible. Then the exposure time is varied until the proper film density is obtained.

Fig. 7—X-ray transmission curves

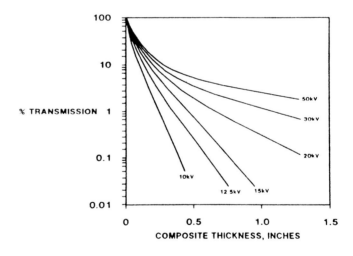

Film Processing

The film processing greatly influences the quality of the radiograph and must be carried out following the manufacturers' instructions. The important processing parameters include concentration of developer and fixator, time of development, amount of radiographs that can be processed with the solution, replenishment, etc. However, a few additional details must be taken into consideration, such as temperature. The importance of the effect of temperature on the development time is shown in Fig. 9. The development-time correction illustrated should be made in order to get the desired density for the chosen exposure.

Fig. 8—*Optimum energy level for different composite thicknesses*

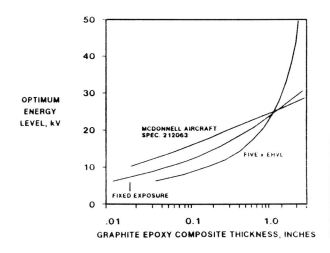

GRAPHITE EPOXY COMPOSITE THICKNESS, INCHES

Fig. 9—*Influence of temperature on the development of radiographs (after Halmshaw³)*

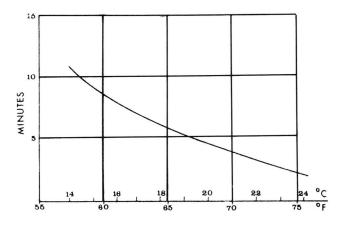

Vertical processing of the radiographs is advised at all times to avoid streaks due to uneven development and fixation. Adequate agitation is required during the development to produce uniform density. The following schedule of agitation is suggested for minimal unevenness of film density: the film should be given a circular motion in its own plane, continually for the first 30 seconds of development, followed by alternate periods of agitation and rest/quiescence for 30 seconds at a time. The use of nitrogen-gas bursts during development may replace part of the agitation; however, one must make sure that this nitrogen agitation really produces an even flow of chemicals across the whole radiograph. Nitrogen gas is used to minimize oxidation of the developer.

After development, the radiograph should be washed thoroughly, either with great agitation or in a stopbath (a 5 percent solution of CH_3COOH), which stops development completely. If this is not performed properly, blotches or streaks may appear on the radiograph due to continued, uneven development.

After fixation, the radiographs are washed in running water for 30 minutes (see Table 2), and then submerged in water with a wetting agent (for 30 seconds) to make the water drain evenly off the film and to facilitate quick and even drying.

TABLE 2—TEMPERATURE DEPENDENCE OF THE WASHING WATER ON WASHING TIME

Temperature	Washing Time
5°-12°	30 min.
13°-25°	20 min.
26°-30°	10 min.

Conclusion

Inhomogeneities in composite materials can be detected rather easily if the defects are not too small. Use of low voltages, as described above, pose some problems not found with classic high-voltage radiography. These problems can be overcome by using suitable film cassette material and by removing lead or fluorescent screens. Experimental determination of the right exposure for obtaining the desired film density proves necessary for high-quality radiographs.

Penetrant-Enhanced X-Ray Radiography for Identifying Matrix Cracks

Introduction

Using low-voltage radiography, overall material properties and large cracks can be observed. However, if one also wants to see small matrix cracks and delaminations in composite laminates, it is necessary to use an X-ray opaque penetrant to make the damage visible. This penetrant fills pores, matrix cracks, delaminations and fiber-matrix debonds and forms a contrast between damaged and undamaged material by absorbing X-rays more readily than the surrounding material. In this section, the choice of penetrant, penetration method and penetration time will be discussed in detail.

The Penetrant

In order to make good use of the penetrant, one must consider the following properties.

(a) The penetrant should be safe to use with normal precautions. Tetrabromoethane, for instance, has a high X-ray absorption and penetrates cracks easily. It is, however, also a potent mutagen and poisonous when inhaled. Many precautions must be taken before one can safely use the penetrant.

(b) The penetrant itself may not induce cracks or debonding in the material. Specifically, thermoplastic matrices can be sensitive to cracking in a solvent environment.

(c) It must have a suitable radio-opacity; it should form a good contrast between the cracks filled with penetrant and the undamaged material.

(d) It must fill the pores and cracks completely, otherwise an inaccurate view of the damage state is obtained.

(e) If possible, it must be removable. If the penetrant still is present when reloading the part, the presence of the penetrant may introduce additional crack growth due to mechanical interaction.

A list of possible penetrants is given in table 3 below (based on Stone).[8]

TABLE 3—RELATIVE RADIOGRAPHIC OPACITY OF PENETRANTS

Organic		Inorganic	
Halogenated Hydrocarbon	Relative Rating	Inorganic Compound	Opacity
Di iodo methane (DIM) Di iodo butane (DIB)	1	Zinc Iodine (ZI)	High
Di bromo methane	2	Silver Nitrate	Medium
Tetra chloro ethylene Tetra chloro ethane	3	Lead Nitrate	Low
Tetra chloro methane Tri chloro ethylene	4	Barium Sulphate	Very Low
Tri chloro methane Di chloro methane	5		
Tri chloro tri fluoro ethane	6		

Fig. 10—Edge penetration using a small load

The criteria for choosing a penetrant depends upon the application and the degree of damage. In cases of gross damage (e.g., ballistic impact), the use of high-contrast penetrants (rating 1-2) will result in losing the fine details.[8] When matrix cracks only are expected, the use of groups 1 and 2 is strongly advised.

The commonly used ZI (supersaturated solution of zinc iodine, isopropanol (10 ml), water (10 ml) and either Agfa Agepon (10 ml)[9] or Kodak PHOTO Flo600 (1 ml)[17] offers the advantages that it is not harmful to human life and the rather inert isopropanol and water offer little risk of producing stress corrosion. However, this penetrant is very corrosive to test equipment and is difficult to remove from the sides of the specimen. Once the penetrant is inside the cracks, it cannot be removed by evaporation, so it may induce damage during further use of the component.

DIM offers a high radio-opacity and good penetration capabilites, and it is volatile, which makes removal of the penetrant from the sides easy. Since it evaporates readily, it leaves the composite part after

Fig. 12—Specimen during penetration

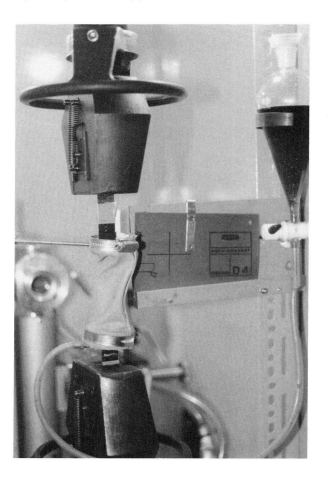

Fig. 13—Specimen during exposure

a few hours to several days. A number of precautions have to be taken, however, because the penetrant is an irritant and can cause skin burns; good ventilation and the use of gloves is advised.

In order to check whether the penetrant induces cracks, it is best to submit the material to a stress-corrosion test. This can be performed by inserting a wedge into a double-cantilever-beam specimen and submerging it in the penetrant. An inspection of crack growth will indicate whether the penetrant attacks the material or not.

Penetration Method

There are two ways of penetrating the specimen with the penetrant: edge penetration[10] or penetration by submersion[11]. The method of penetration depends mainly on the size of the composite part to be examined. If the part is small, then the submersion technique will prove to be the best. If, however, the composite structure is large, one will have to be content with edge penetration.

Edge Penetration

In edge penetration, the specimen is mounted on a horizontal tensile testing machine and is given a small load (usually 5 percent of the failure load). The penetrant is applied manually drop by drop with a syringe to the sides for 30 minutes.

The advantge of this method is that it is easy to perform on any kind of structure. The disadvantages are as follows: (1) Longitudinal cracks can only be penetrated through transverse matrix cracks. (2) It is difficult to perform *in situ* on a tensile testing machine since the penetrant drops applied to the edges tend to glide down (vertically). (3) A small load in comparison to the earlier applied load must be used, to avoid introducing new damage during loading. (4) The penetrant is usually only applied to the sides and not to all the faces of the specimen. (5) The penetration is time consuming and difficult to reproduce since it is performed manually. Experiments have shown that a penetration time of 30 minutes is sometimes insufficient to penetrate the cracks to their full extent.

Submersion Penetration

In submersion penetration, the specimen is either submerged during loading or submerged into the penetrant without any load. The advantages include: (1) excellent control over affected region, amount of penetrant (completely surrounded) and penetration time, and (2) better penetration, since the specimen is continuously, and if necessary for a long time, surrounded by the penetrant. The disadvantages include: (1) Long-term exposure of the composite part to the penetrant is necessary because most of the cracks are closed. This may produce problems when using a corrosive penetrant. (2) A large amount of (expensive) penetrant may be necessary to submerge the part.

Sometimes it is not possible to submerge the part in the penetrant. A solution may be to use a volatile penetrant and to expose the part to the penetrant gases. Experiments have shown that gaseous di-iodo-methane penetrates large matrix cracks easily. Small cracks have to be exposed for a longer time to fill the cracks completely.

In-Situ Radiography During Monotonic or Fatigue Loading in Combination with Submersion Penetration

When radiography is being used in the laboratory for studying damage development in composite materials, *in situ* radiography proves to be an excellent method for studying matrix crack growth during monotonic or cyclic loading.[12] The X-ray equipment, together with an X-ray shield (to keep the X-rays confined to the test region) is mounted on the

Fig. 14(a)—(9_2 90_2)$_s$ graphite epoxy specimen after a tensile test; Penetration time: 60 seconds in DIM

Fig. 14(b)—Same specimen; penetration time: 60 minutes in DIM

Fig. 15—Partial penetration using the submersion technique

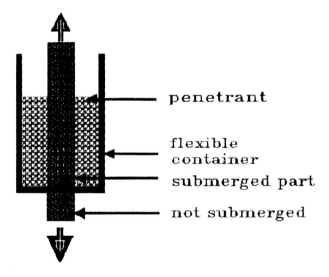

penetrant

flexible container

submerged part

not submerged

tensile/fatigue testing machine. This *in situ* radiography has the advantage that many radiographs can be taken of the changing damage state during tensile or fatigue testing without having to remove the specimen from the machine. At the same time it is possible to radiograph the matrix cracks when they are opened. Penetration is accomplished by a simple penetration device (Figs. 12 and 13) which consists of a flexible rubber tube which can be raised (to penetrate the specimen) or lowered (to radiograph the specimen).

Damage Documentation Using Penetrant-Enhanced X-ray Radiography

Accurate damage documentation is only possible when the penetrant fully penetrates the existing

damage and if the penetrant doesn't induce crack growth. How good penetration is obtained and how one can find out whether a penetrant induces cracks will be discussed below.

The extent of penetration is dependent on the surface tension of the penetrant and the penetration time. The penetration time depends mainly on the size of the part and the size of the damage in the material. To determine the penetration time, incremental penetration is used. Here we penetrate the part for a certain length of time, radiograph it, and penetrate it again for a certain length of time, after which another radiograph is taken. The radiographs are compared for the extent of damage. This process is repeated until no additional damage can be observed. Figure 14 shows two radiographs taken after different penetration times.

Influence of the penetrant on the damage development. To determine whether the penetrant influences the crack growth, the following steps are taken. First of all, the sides of a polished specimen are meticulously inspected for matrix cracks and debonding. After this inspection, the specimen is submerged without load for one day. The same visual inspection is repeated and a radiograph is taken. If there is additional damage, the penetrant should not be used for damage inspection. Additionally, after visual inspection, the specimen is partly penetrated while under load (see Fig. 15). After 24 hours of partial penetration, the specimen is fully submerged in the penetrant for another 24 hours.

After this penetration, the specimen is radiographed and again submerged for 24 hours without load and radiographed. This enables one to compare the two parts, one penetrated for 24 hours under load and 24 hours without any load, and the other part penetrated for 48 hours without any load. The radiographs and the visually observed damage are compared; if they indicate a difference, the penetrant is attacking either the interface or the matrix.

Typical X-Ray Conditions

For one to determine the damage state (e.g., matrix cracks, delaminations) of a composite part,

TABLE 4—TYPICAL X-RAY CONDITIONS FOR CARBON EPOXY LAMINATES, 1 MM THICK	
Characteristic	value
film to focus distance	50 cm
Voltage	16 kV
filament current	9mA
exposure time	1.0 minutes
screens	Pb at back side only
Film speed	100 (eg AFGA GEVAERT Structurix, D4)
Cassette	plastic (eg AGFA GEVAERT vacupac D4)

penetration is necessary. For a carbon epoxy, the penetrant di-iodo-methane is advised. The part is submerged unloaded in the penetrant for 24 hours and exposed to X rays. The X-ray conditions are given in Table 4.

Radiographic Definition

The problem of radiographic definition can be described in terms of lack of image sharpness or blurriness, and image contrast. The effective blurriness can be calculated as the result of several factors: lack of geometric, screen and film sharpness. The magnitude of any of these can vary over a wide range, depending on the equipment, techniques and specimen thickness. The effect of lack of radiographic sharpness is a blurred image and, also, reduction of the contrast obtained in the image of a small detail. Since the defects in composite materials are mainly of a very small scale, this problem should be taken into consideration.

Geometric unsharpness. Due to the finite size of the X-ray spot, every point of the object has several projections on the X-ray film (Fig. 16). By simple geometry one can express the geometric unsharpness as:

$$U_g = \frac{s \times b}{a} \qquad (1)$$

The magnitude of geometric unsharpness can vary markedly. The majority of industrial X-ray tubes have focal spots of 0.1-5 mm in diameter. Table 5 below (based on Halmshaw), shows some typical values for U_g. This unsharpness increases the width of the projection of the defect and reduces image contrast.[3]

TABLE 5—TYPICAL VALUES OF GEOMETRIC UNSHARPNESS FOR A RANGE OF INDUSTRIAL RADIOGRAPHIC TECHNIQUES				
Radiation	Source	Specimen	sfd	Ug
100kV X-rays	5	6	500	0.06
200kV X-rays	5	25	1000	0.13
400kV X-rays	7	75	1000	0.57
Co-60 τ rays	4	50	500	0.44

Film unsharpness. A photographic emulsion consists of silver-halide grains which are suspended in an emulsion which is coated on a plastic support. When a rontgen quantum is absorbed in such a silver-halide grain, it can then be developed, and a photo electron is released. When this photo electron has enough energy, it may be absorbed by an adjacent halide grain which itself can then be developed. The more energy such a photo electron has, the larger the range it has and the more grains that will be influenced (e.g., at a 1000 kV radiation ±80 grains are affected).[5] As the effect of film unsharpness is to produce a small area of developable halide grains around a theoretically sharp point of X rays, the result is the blurring of the image. The magnitude of film blurriness depends on how closely packed the silver-halide grains are: a closer packing reduces the film blurriness because the photo electrons will travel a very short range before they are absorbed by the adjacent grains. Fortunately, for voltages up to 33 kV, it has been shown, only one halide grain is in-

Fig. 16—Geometric unsharpness, U_g as it affects the image of a small defect (after Halshaw[3])

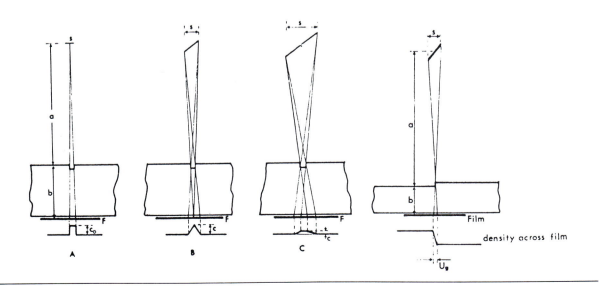

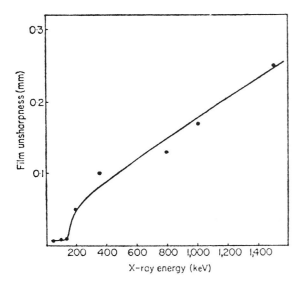

Fig. 17—Film unsharpness as a function of X-ray energy

fluenced by an incoming X-ray quantum; thus, when radiographing composite materials, this problem does not arise.[13]

Screen unsharpness. The loss of definition resulting from the use of salt-intensifying screens is ascribed to the image spread caused by the reflection and scattering of the light emitted by the crystals of the fluorescent material on absorbing X-ray quanta. To reduce this screen blurriness, the screen is put as close as possible to the film. Since the intensifying effect is rather low when using low voltages (Table 6, based on Fassbender and Hagemaier), one can simply leave out these screens and thus eliminate the problem.[14]

Fig. 18—Two different exposure times of the same radiograph. It is evident that one has to be careful in interpreting photographic enlargements

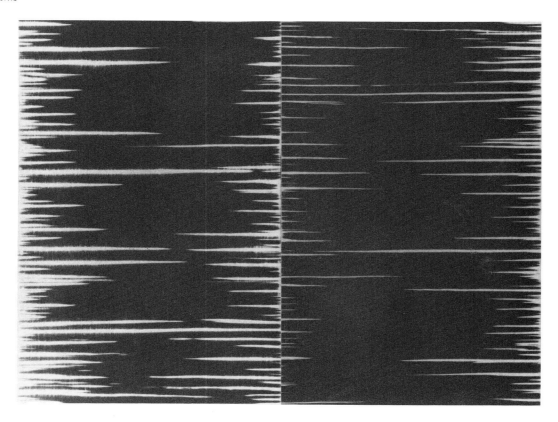

TABLE 6—EXPOSURE TIME REDUCTION DUE TO THE USE OF FLUORESCENT SALT SCREENS USING KYOKKO FLUOREMETALLIC SCREENS		
Voltage (kV)	Screen	% Reduction in Exposure
10	SMP 308	
20		33
30		38
40		39

Image-Quality Indicators

The quality of the radiographic technique is assessed by the use of penetrometers or image-quality indicators (IQI). For radiography on metals, such IQI are available. However, for composite materials, one must make his or her own penetrometers. A thin laminate (0.5 mm thick) which has been penetrated and shows matrix cracks can serve as a good penetrameter. This laminate with known damage is placed on top of the composite part and radiographed together with the composite part. This penetrometer will indicate whether the exposure of the composite was chosen well.

Fig. 19—Sketch showing the orientation of specimen and film relative to X-ray source and the resulting relative location of artifacts on the film (after Sendeckyj[15])

Conclusions

Penetrant-enhanced X-ray radiography proves to be necessary when one is interested in very small details such as matrix cracks or in larger defects with a bad orientation with respect to the X-ray source (e.g., delaminations). The choice of penetrant and the method of penetration is important for good documentation of the damage present in the material.

Special Techniques

Radiographic Enlargement

It is evident that when the features we are interested in are of a small scale, some degree of enlargement will be necessary. Either, micro-radiography (described below), photographic enlargement of contact radiographs or a microfiche reader (enlargement of 10-50 X) is used. When one is interested in very fine details, a microscope may prove to give additional information. The use of very fine-grain film is strongly advised in this case. When using photographic enlargement, one must be sure that the film-exposure time is chosen well; underexposure of the photograph sometimes blurs fine details which are visible on the radiograph.

Stereo Radiography

It is sometimes desirable to be able to determine the exact position and size of a defect in a two- or three-dimensional part. Since it is rarely possible to radiograph from two perpendicular directions, and computer aided tomography is usually not available and too expensive for the job, stereo microscopy is, in most cases, a good solution to the problem.[15-17] Stereo X-ray radiography is the radiographic equivalent of optical stereography, in which two images are produced from two slightly different angles and then are optically recombined to produce an apparently three-dimensional view.

Microradiography

The small diameter of all fibers used in composites and the related size of damage suggest that a point source of X-rays be used. Projection-radiographic techniques can be used with advantage to improve resolution. In order to reduce the accompanying image blurriness, one must reduce the focal spot drastically. Typically, a 1- to 5-micron focal spot is used along with a magnification of about 100. Thus, failure of carbon fibers in thin sections can be revealed, as well as very fine distributions of subsurface cracks in damage zones near crack tips.[18]

If primary magnification is not required, the source-to-film distance can be significantly reduced without a loss of image definition. This means that the exposure time can be reduced and also that radiographs can be obtained in cases where access to the area to be inspected is limited. Small-diameter (4 mm) microfocus rod anodes may be very helpful in these cases.

Neutron Radiography

While the X-ray mass-absorption coefficient increases monotonically with atomic number of the constituents, the neutron-absorption coefficient does

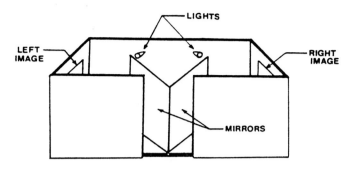

Fig. 20—Device for viewing X-ray stereo pairs, after Sendeckyj[15]

Characteristic curve: the plot of density vs. log of exposure or of relative exposure.

Subject contrast: the ratio (or its logarithm) of the radiation intensities transmitted by selected portions of the specimen.

Radiographic contrast: the difference in density between an image and its immediate surroundings on a radiograph.

Film contrast: a qualitative expression of the slope of the characteristic curve of a film; that property of a photographic material which is related to the magnitude of the density difference resulting from a given exposure difference.

not and is significantly higher for selective elements, such as hydrogen.[19] This characteristic behavior makes neutron radiography suitable to detect variations in hydrogen content and thus allows us to observe resin-rich zones.

Martin[20] has calculated the absorption coefficients for carbon epoxy with a 28 percent fiber content and found a 0.7 percent change of X-ray absorption and a 2.6 percent change of neutron absorption for a one-percent fiber-content change. X-rays cannot measure small changes in resin content, whereas neutron-gauging techniques are capable of measuring resin-content changes of one percent.

One disadvantage of neutron radiography is that the neutron sources are not as accessible as those of X-rays. The presence of water in laminate after environmental exposure can be detected also.

Appendix

Definitions

Exposure: Radiographic exposure is commonly expressed in terms of milli ampere seconds or milli curie hours for a known film to source distance.

Density: the quantitative measure of film blackening,

$$D = \log I_o/I$$

where

D = density,
I_o = light intensity incident on the film
I = light intensity transmitted

References

1. Chang, F.H., Couchman, J.C., Eisenmann, J.R. and Yee, B.G.W., "Application of a Special X-Ray Nondestructive Testing Technique for Monitoring Damage Zone Growth in Composite Laminates," Composite Laminate, ASTM STP 580, ASTM, Philadelphia, PA, 176-190 (1974).

2. Jamison, R.D., "The Role of Micro Damage in Tensile Failure of Graphite/Epoxy Laminates," Comp. Sci. and Tech., 24, 83-99 (1985).

3. Halmshaw, R., Industrial Radiology: Theory and Practice, Applied Science Publishers, London (1982).

4. Huggins, B.E., "Radiographing with Low Energy Radiation," Brit. J. of NDT, 119-125 (May 1981).

5. Crane, R.L., Chang, F. and Allinokov, S., "The Use of Radiographically Opaque Fibers to Aid the Inspection of Composites," Mat. Evaluation, 69-71 (Sept. 1987).

6. Scott, I.G. and Scala, C.M., "A Review of Nondestructive Testing of Composite Materials, NDT Int., 75-86 (April 1982).

7. Wysnewski, R.E., "Revelations at Low kV," Pyne Corp., Larchmont, NY.

8. Stone, D.E.W. and Clark, B., "Nondestructive Evaluation of Composite Structures—an Overview," Proc. European Symp. on Damage Development and Failure Processes in Comp. Mat., Leuven, Belgium, ed. Verpoest, Wevers (May 4-6, 1987).

9. Schulte, K., Henneke, E.G. and Duke, J.C., "Methoden zur Untersuchung des Schadigungsverlauf von CFK bei Ermudingsbelastung, Verbundwerkstoffe-Stoffverbunde, Vortrags—und Discussionstagung von 9 bis 11 mai 1984, Konstanz, Deutsche Gesellschaft fur Metallkunde, Oberursel (1984).

10. Sendeckyj, G.P., Maddux, G.E. and Tracy, N.A., "Comparison of Holographic, Radiographic, and Ultrasonic Techniques for Damage Detection in Composite Materials," ICCM2, Proc. 1987 Int. Conf. on Comp. Mat., ed. B. Noton, R. Signorelli, K. Street and L. Phillps (1987).

11. Van Daele, R., Verpoest, I. and De Meester, P., "In Situ Radiography as a Means of Calibrating Acoustic Emission," Composites Evaluation, Proc. 2nd Int. Conf. on Testing, Evaluation and Quality Control of Composites-TECQ, Univ. of Surrey, Guildford, UK, 117-127 (Sept. 22-27, 1987).

12. Van Daele, R., Verpoest, I. and De Meester, P., "Matrix Cracking in Cross Plied Thermosetting and Thermoplastic Composites During Monotonic Tensile Loading," Advancing with Composites, Int. Conf. on Comp. Mat., Milan, Italy, 1, 143-156 (May 10-12, 1988).

13. Halmshaw, R., "Physics of Industrial Radiology", ed. R. Halmshaw, American Elsevier Publishing Co. Inc., New York, N.Y., 178 (1966).

14. Fassbender, R.H. and Hagemaier, D.J., Low Kilovoltage Radiography of Composites, Mat. Evaluation, **41**, 831-838 (1983).

15. Sendeckyj, G.P., Maddux, G.E. and Porter, E., "Damage Documentation in Composites by Stereo Radiography," Damage in Composite Materials, ASTM STP 775, ed. K.L. Reifsnider, ASTM, 16-26 (1982).

16. Stone, D.E.W., "The Use of Radiography in the Nondestructive Testing of Composite Materials", RAE Tech. Rep. TR 71235, 31p (Dec. 1971).

17. Blom, A.F. and Gradin, A., "Use of Radiography for Nondestructive Testing and Evaluation of Fiber Reinforced Composites," Flygtekniska Forsoksanstalten, FFA TN 1985-53.

18. Harris, B., Assessment of Structural Integrity of Composites by Nondestructive Methods, Rep. of the School of Material Science, Univ. of Bath.

19. Bar-Cohen, Y., NDE of Fiber Reinforced Composite Materials—a Review, Mat. Evaluation, **44**, 446-454 (March 1986).

20. Martin, B.G., "An Analysis of Radiographic Techniques for Measuring Resin Content in Graphite Fiber Reinforced Epoxy Resin Composites," Mat. Evaluation, 65-68 (Sept. 1977).

Section VIIA

Vibrothermography Applied to Polymer Matrix Composites

by Edmund G. Henneke II

Introduction

Vibrothermography is a technique that combines mechanical, vibrational excitation with real-time video thermography to detect defects in advanced composite materials. Basically, a mechanical oscillatory load is applied to the material. Mechanical energy is dissipated as thermal energy throughout the material, but preferentially in regions surrounding delaminations. The resulting temperature distribution is detected by monitoring the infrared radiation emitted from the surface of the specimen with a real-time video-thermographic camera. Regions that have near-surface delaminations are easily detected by the temperature-gradient pattern that is established.

The vibrothermography technique has been applied to polymer-matrix and metal-matrix composites. As this technique is most sensitive, however, to delaminations or delamination-type defects in polymer matrix composites, this chapter will emphasize the applications and selected results that have been obtained for this material type.

Thermography

Thermography is the measurement and graphing of isothermal contours on the surface of an object. Any method which is capable of measuring temperature spatially can be used to produce a thermograph. However, the method preferred because of its many advantages is real-time video thermography. All matter at temperatures above absolute zero spontaneously emits electromagnetic energy at wavelengths longer than those corresponding to the red portion of the optical spectrum. The frequency range covered by the infrared spectrum is shown in Fig. 1. This 'infrared' emission occurs because of the thermal motion of subatomic particles, atoms and molecules. The following sections present a brief overview of the physics of infrared radiation and its detection by real-time video-thermographic cameras.

Infrared Radiation

Because of the wide range of energies possessed by the various types of motions of the basic building

Fig. 1—The electromagnetic spectrum

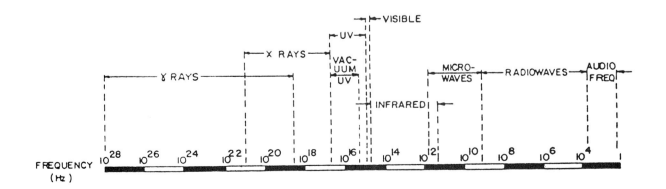

blocks of matter, and the quantum jumps available to the basic particles, the infrared spectrum itself covers a broad range of frequencies. Classically, the infrared spectrum was categorized into three subregions—near, intermediate, and far infrared, depending upon the value of the infrared frequency in relation to the red spectrum. The initial reason for this classification was due to the different experimental techniques required to detect the radiation in these different bands. However, in general, one can discuss these three bands in relation to the mass of the source of the infrared: (1) subatomic particles whose quantum jumps are responsible for the majority of near infrared, (2) atomic particles whose thermal vibrations produce the majority of intermediate infrared, and (3) molecules whose vibration and rotation motions produce the far infrared.[1]

The thermal energy emitted by a solid object depends upon the properties of its surface a well as its temperature. In particular, the emitted energy depends upon a parameter known as the thermal emissivity[2], e. The thermal emissivity is defined as the ratio of infrared energy emitted from a point on the surface to the energy that would be emitted from a point having the same temperature on an ideally emitting body. An ideally emitting body is called a black body and obviously has an emissivity of one. Ideal black bodies can be approximated in the laboratory by measuring the infrared energy emitted through a small hole in a body with an internal cavity which is at constant temperature. (See, for example, Ref. 1 or 2.) Figure 2 presents schematics of cavities which can be used to approximate black-body radiators.

The analytical model which predicts the thermal-radiation energy emitted by a body was established by Planck. With his quantum hypothesis, Planck suggested that a harmonic oscillator (an atomic or other particle moving in a linear potential field may be thought of as an harmonic oscillator), moving in one dimension, could not possess *any* value of energy in a continuum of energies, but could have only a total energy value which would satisfy the relationship

$$E = nh\theta \qquad (1)$$

where E is the total energy of the oscillator, θ is the frequency of the oscillator, and h is a universal constant (Planck's constant, h = 6.625 x 10^{-34} W^2. Because of the discrete nature of the atoms or molecules composing a mass of matter, when one calculates the total energy contained therein, one must use an infinite discrete summation, in comparison with continuous energies which require a continuous integration for total energy calculation. The difference in mathematical properties of a discrete summation in comparison with a continuous integration led Planck to derive his well-known distribution law for spectral-radiation emittance:

$$W_\lambda = 2\pi e \,(10^{-9})\, hc^2/\lambda \{ \exp(hc/\lambda kT) - 1\} \qquad (2)$$

where W_λ is the intensity of radiation of wavelength, λ, emitted ($W/m^3/m\mu$), c is the speed of light (2.99793 x $10^2 m/s$), exp is the Naperian or natural base of logarithms, k is Boltzman's constant (k = 1.3804 x $10^{-23} J/°K$), and T is the absolute temperature (°K).[2] Examples of the spectral-radiation curves for different temperatures are shown in Fig. 3 for a black body.

Fig. 2—Schematic forms of cavities useful for simulating black-body radiators

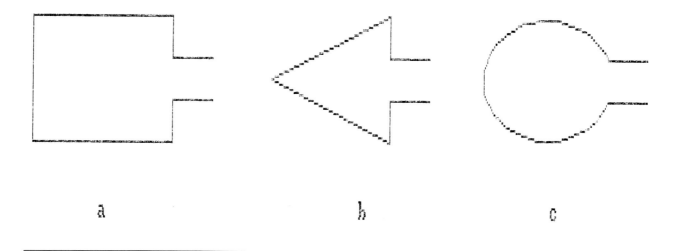

a   b   c

130

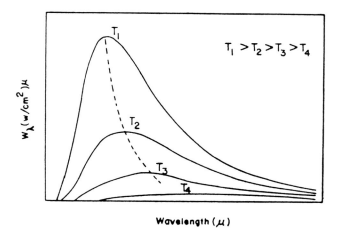

Fig. 3—Spectral radiance curves for a black-body radiator

$$T_1 > T_2 > T_3 > T_4$$

Wavelength (μ)

The parameter, e, is the emissivity of the surface. The emissivity may be a 'total' emissivity—the fraction of total energy emitted at a given temperature—or a 'spectral' emissivity—the fraction of energy emitted at a particular wavelength (and, of course, temperature). The total emissivity of a body might be quite low while the spectral emissivity for certain wavelengths may approach unity. Both types of emissivity may vary with such factors as temperature, physical state, surface finish, molecular surface layers, etc. The emissivity cannot be calculated from any basic physical model but must be determined experimentally for each body (and often for different points on the surface of the same body). It is strictly a surface characteristic for opaque materials. The emissivity ranges from zero for 'mirror-like' surface to nearly one for lamp black, zapon black, and such surfaces. Interestingly, human skin has an emissivity of nearly one.

Because of the variations in the emissivity of a surface, the determination of an absolute value of temperature by measurement of the intensity of emitted infrared radiation is fraught with difficulties. It is possible, for example, that a particular surface will have a sufficiently large reflection coefficient that the body heat emitted by an experimenter might reflect off the surface of the examined object and be interpreted as a high temperature for that object or area of reflection. One can reduce difficulties such as this, or surface variations of emissivity, when studying composite materials, by painting the surface to be studied with a uniform coating of lamp black, for example. Some users have even found success using powder-spray deodorant as a surface covering. This has the advantage of being easily removed from the surface and also has an emissivity nearly equal to one. However, there is sometimes a difficulty with the powder remaining on the specimen when the specimen is shaken for vibrothermography. While the difficulty involved with determining an absolute-temperature value will still be present after coating the material, the uniform surface coating will allow one to have some confidence that the temperature gradients observed on the surface are a result of material temperature differences and not spurious reflections from hotter objects in the surrounding environment, nor local surface emissivity differences.

Other important physical characteristics of infrared radiation are of interest to the experimenter. Figure 3 presents a schematic representation of a family of spectral radiance curves for radiation from a black body at several different absolute temperatures. Such curves were determined experimentally before Planck's mathematical law was found to quantitatively predict their form so closely. One first notes that black-body radiation is a smooth, continuous function of wavelength with a single maximum occurring at one value of wavelength, λ_m. This observation led to the statement of Wien's law:

$$\lambda_m \doteq b \, / \, T \qquad (3)$$

That is, the wavelength, in microns, at which the maximum intensity of radiation is emitted is inversely proportional to the absolute temperature, °K. The constant of proportionality, b, is known as the Wien displacement constant and has a value of 2897 microns/°K. Physically, the Wien law states that the higher the temperature, the more the peak of radiation shifts towards shorter wavelengths and, hence, higher frequency and higher energy content [eq. (1)]. The progression towards smaller wavelengths for increasing temperature is a phenomenon which most people have seen at one time. At moderate temperatures, the spectral-emission peak is at wavelengths in the infrared and invisible to the human eye. As the temperature increases, the object will gradually take on a reddish hue, i.e., the wavelengths of emitted infrared radiation shorten to include the beginning of the red band. At yet higher temperatures, the red brightens, changes into orange, yellow, and finally white, when the emission spectrum extends to cover the green and blue bands. Perhaps the status of the earliest practitioners of thermography must be given to blacksmiths who quite early learned to gauge the proper tempering and heat-treating temperatures of steel by judging the color to which the alloy was heated.

A second important observation which can be made from Fig. 3 is that there is a single emission curve for each temperature of the black body. Furthermore, no emission curve ever intersects another. In particular, each curve lies above all other curves corresponding to lower temperatures of the black body. This fact means that if proper techniques are developed for measuring the spectral-emission curves, one can uniquely determine the absolute temperature of a black body. Recall, however, that because of emissivity variations for real emitting bodies, this absolute-temperature determination offers many difficulties to the experimenter.

Detection of Infrared Radiation by Video Thermography

Two basic types of infrared detectors are in use: (1) photon-effect devices and (2) thermal devices. The photon-effect devices are sensitive to the wavelength of the received radiation while thermal devices respond only to the degree of heating caused by the incident radiation and are largely independent of wavelength. The performance of real-time thermography, such as with a video-thermographic system, requires that the entire field of view be scanned very rapidly so that the temperature of each field point can be measured and displayed many times each second. Such systems require the high sensitivity and very rapid response time of photon-effect devices.

Photon-effect devices utilize solid-state materials which produce voltage, current or resistance changes when irradiated by photons. These semiconductor materials are generally classified as photoemissive, photoconductive, or photovoltaic detectors[1]. Because of their faster response times, photoconductive or photovoltaic devices are more appropriate for use in video-thermographic cameras. For maximum sensitivity, and to reduce extraneous thermal noise, it is usually necessary to cool the semiconductor material to low temperatures. Most commercially available thermographic-detection systems require liquid nitrogen be used for this purpose. Some more recently available equipment has thermoelectric-cooling devices installed to reduce the detector temperature.

Real-time video-thermographic systems perform a fast scan on the test surface via a complex mirror

Fig. 4—Schematic of internal mirror arrangement for a thermographic camera (after Bergstrom and Borg[3])

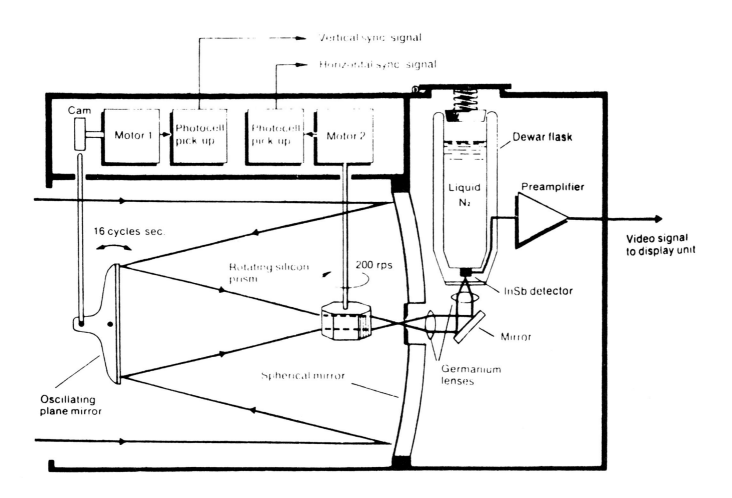

system which rotates at a very high speed. The infrared radiation emitted from a small region of the surface is reflected by the mirrors onto the semiconductor sensor (Fig. 4[3]). As the mirrors rotate, a complete picture of the scanned surface is built up. The response of the thermographic detector to the received infrared radiation is conditioned and displayed upon cathode-ray monitors at rates approximating television-display rates in either gray scale or color format. The gray-scale monitors display continuous temperature readings via shades of gray between black and white. The color-display monitors select a series of distinct colors (typically 10) to display those regions of temperature which lie in a window, or band, of temperature equivalent to the total temperature range divided by the number of colors available (ΔT = 1/10th the temperature range for displays having 10 colors). With these devices, one can monitor simply and rapidly the temperature-gradient profiles on the surface of the test objects for total temperature ranges between the order of 1 and 1,000° C, depending upon the particular instrument. Specific information and details on instrument capabilities can be obtained from individual suppliers.

Two additional considerations are of some importance to the application of real-time video thermography to vibrothermography: the system geometric relationships and the fidelity of surface thermal patterns to subsurface flaws. For an infrared system viewing any source, the received power at the system aperture is given by

$$H = W_\lambda \omega / \Omega \qquad (4)$$

where W_λ is the spectral emittance, given by eq. (2), ω is the angular field of view of the viewing system (defined by the optical system and the detector) and Ω is the total solid angle about the source. Thus, it can be seen that the primary geometric parameter which governs the response of the infrared viewing system is the relationship between the angular field of view of the system and the angle subtended by the source. If the source is small compared with the field of view of the detector (that is, a point source), the received radiation will vary with the distance between the source and detector but not with the angle about the source. On the other hand, if the source is large compared with the field of view, the received radiation varies with neither the distance to the source nor the angle about it. This latter fact is a result of Lambert's cosine law which states that the radiant energy emanating in a given direction from any point on a surface is a function of the cosine of the angle between the normal to the surface at that point and the given direction. As a consequence, the maximum radiation occurs along the normal direction to the surface and none tangentially to it.

The significance of Lambert's law for infrared detection is that a detector viewing an emitting surface will always receive the same amount of energy no matter what the angle between the detector's line of sight and the normal to the radiating surface, if the source is large compared to the angular field of view of the detector. This statement is true for both planar and general curved surfaces, although it is easier to see and understand for planar ones. As the angle between the line of sight and the surface normal increases, so does the surface area viewed by the detector (in fact, the viewed area increases inversely with the cosine of the angle between the line of sight and the surface normal). The increase in viewed area exactly matches the reduction in radiation as given by Lambert's cosine law. Hence the total energy incident on the detector is constant (also assuming that the temperature of the viewed surface is uniform). This, of course, is valid only as long as the emitting surface completely fills the detector's field of view and therefore obeys the assumption of a large source compared to the viewing angle. A simple rule in utilizing a detector system then is to make the source the only object in the field of view by appropriately controlling the field stop of the viewing system.

The second consideration of importance to understanding the application of vibrothermography to nondestructive evaluation of materials is the fidelity of the observed surface thermal patterns to the interior inhomogeneities or flaws. The relationship between the surface isotherms and the interior thermal patterns is, of course, governed by the thermal conductivity of the material and the distance between the surface and an interior region of interest. For very thin materials such as many composite laminates, it has been shown that the fidelity of surface thermal patterns to interior flaws is quite good.[4,5] For bulk materials, thermography may be useful for qualitative indications of the presence of flaws but will not be nearly as good as other NDE methods for resolution of the flaw size and shape. For example, what might seem to be a rather sharp discontinuity may prove to have a weak thermal signature because of heat conduction through the material.[6] The thermal signature can be improved by making the cooling rate at the surface as large as possible. This might be done by making the surface as nonreflective as possible and by augmenting radiant cooling by forced air circulation over the test surface.

Vibrothermography

Vibrothermography is the combined application of mechanical excitation and real-time video thermography to the nondestructive evaluation of materials, Fig. 5. The mechanical excitation is used to establish steady-state strain patterns in the material. Irreversible transformation of the vibrational mechanical energy into thermal energy occurs, thereby producing heat patterns throughout the specimen (Fig. 6[7-13]). The surface heat is, of course, partially irradiated as infrared electromagnetic waves, which can be detected by a thermographic camera. If flaws such as delaminations exist in the

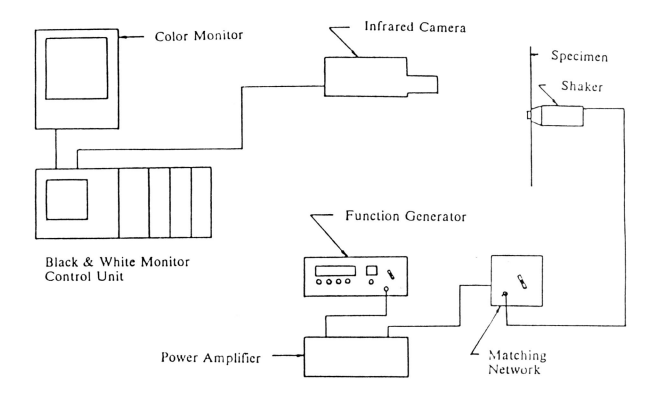

specimen, regions of strain concentration around the flaws will produce more heating than homogenous, unflawed regions of the specimen[.7] This increase in the heat produced will evidence itself as temperature gradients in the thermograph. Figure 7 presents a sample thermograph of a graphite-epoxy specimen with an internal delamination.

Techniques other than the application of mechanical excitation for the development of heat patterns have been employed[.14-16] If the user is especially interested in the mechanical properties of the material he or she is examining, the application of mechanical-vibrational energy has a particular advantage. Because the development of heat during the vibrothermography examination is directly related to the strain field in the specimen, the thermograph is directly related to the mechanical condition of the material. Hence, the vibrothermographic technique is better able to describe the mechanical state than, for example, is a thermographic procedure which uses external heating of the specimen to produce temperature patterns through differences in thermal conductivity from region to region in the specimen. Procedures for producing mechanical excitation in the material for the purpose of performing a vibrothermographic examination are described in the next section.

Mechanical Excitation

Mechanical excitation of a material for vibrothermography can be generally categorized into low-frequency/high-amplitude or high-frequency/low-amplitude. Low-frequency/high amplitude testing is generally performed in conjunction with a fatigue test where the high-amplitude, cyclic, mechanical-fatigue loads provide the source of mechanical energy[.7] This loading situation may *not* be strictly nondestructive, as the level of loading may be such that fatigue damage is occurring in the specimen. However, if the fatigue loading is being applied anyway, for test purposes or as a natural consequence of in-service application of the material, thermographs can be taken of the material surface. An example is shown in Fig. 6. The performance of thermography itself is obviously nondestructive. High-frequency/low-amplitude vibrothermography testing more exactly fulfills the definition of nondestructive testing. High-frequency/low-amplitude testing is normally performed by attaching the material to be examined to a mechanical shaker which can inertially load the material, as described subsequently.

Active heat generation in composites as a result of the transformation of mechanical into thermal

Fig. 6—Example thermograph of specimen showing changing temperature patterns developed during fatigue

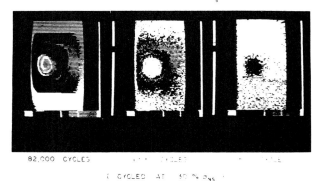

FATIGUE DAMAGE DEVELOPMENT IN A [±45,0]$_s$ FLAWED LAMINATE

82,000 CYCLES

(CYCLED AT 50 % σ_{NS})

(a) 500 cycles, (b) 1590 cycles, (c) 82000 cycles

energy will be discussed in three categories. First, it is well known that when most materials are cyclically deformed, even in the elastic range, some of the mechanical energy is dissipated by nonconservative micromechanical deformation processes such as dislocation motion, impurity diffusion, and other complex local molecular or atomic activity. For

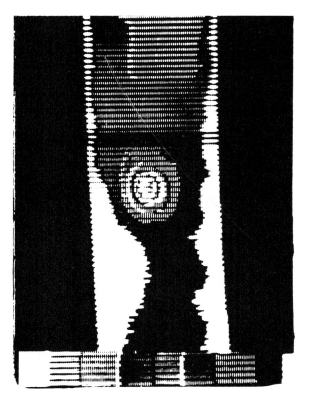

Figure 7—Example thermograph of delamination in graphite-epoxy laminate vibrated at resonant frequency of local damaged region

polymeric materials, these processes bring about what is commonly called viscoelastic response. The amount of energy dissipated by such mechanisms may range from imperceptible amounts up to several percent of the input energy. Most of the attention that has been given to hysteresis has centered on crystalline solids. In this case, anelasticity of the material is concerned with atomic diffusion events that are activated by stress, including grain-boundary motion, single- or paired-solution atom motion, and twin-boundary activity. These events are prominent in the small strain range, but the amount of heating produced by them is not significant because of the low ratios of dissipated-to-input energy, a low level of input energy for small strains, and the low frequencies of oscillation at which these events are commonly excited. For composite materials with polymeric matrices, viscoelastic dissipative hysteresis dominates the heat generation, but hysteresis-energy heat patterns can be generated in metal-matrix composites as well.[13]

Two parameters are especially important in the generation of such heat patterns using mechanical excitation, as is indicated in Fig. 8(a). The first of these is the stress or strain level at which the specimen is deformed. Depending upon the material used, the deformation process, the type of defect, and other variables, a stress or strain level greater than or equal to roughly one third of the level required to fail the specimen is sufficient to produce observable heat patterns around defects of engineering interest in most cases. The second major parameter of importance is the cyclic frequency with which the mechanical load is applied. If other things are equal, the amount of power introduced into the specimen is directly proportional to the frequency of excitation, so that the optimum applied frequency is the highest frequency available. This is true since temperature differences in the region of a defect will be proportional to the differential amount of energy generated by the defected and adjacent, undefected regions. It should be mentioned, however, that energy-dissipation mechanisms may be frequency and temperature dependent so that the optimum frequency may not be the largest one easily obtainable with the test rig available. Trial and error is required. Most heat patterns that are generated in composite materials using this active hysteresis-energy-emission scheme are created by cyclic frequencies between approximately five and 30 Hz. As already pointed out, one of the advantages of the mechanical excitation scheme is the fact that the dissipation of the defect or defected region is in direct proportion to its mechanical importance, or in another sense, to its mechanical disturbance of the materials response in that region. Hence, an element of interpretation is added to the information obtained in this way. It should also be mentioned that while this procedure can be used nondestructively, it is also possible to watch the growth and development of defects by simply observing the cyclic application of load levels and load

135

histories which are sufficient to cause such growth. In that way, the chronology of damage development can be followed.

Figure 8—Examples of mechanical excitation for heat generation

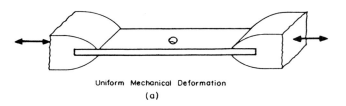

Uniform Mechanical Deformation
(a)

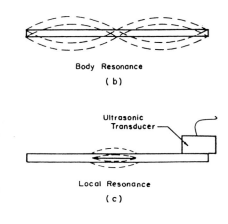

Body Resonance
(b)

Ultrasonic Transducer

Local Resonance
(c)

Figure 8(b) shows another example of an active heat-generation procedure. If proper frequencies of mechanical excitation are chosen, resonant body vibrations of the specimen or component can be excited. Energy is dissipated, and a heat-emission pattern is formed, in proportion to the stress distributions which are created by the resonance-vibration mode. Defects and flaws, or damaged regions, are revealed not only by the manner in which they disturb the generation of the heat pattern, but also by the manner in which they disturb the vibration-mode shape as revealed by the heat pattern (Fig. 9).

Figure 8(c) indicates yet another method of mechanical excitation. Higher frequencies, introduced into the specimen by a wide-band mechanical shaker, are varied over a frequency range until local flawed regions in the material are set into local resonance.[1-13] In addition to the hysteresis-energy dissipation in the region of local vibration, another very strong source of energy dissipation frequently develops due to the strain concentration around the region occupied by the flaw. In this application, vibrothermography has the capability of selectively resolving different sized defects by the use of different frequencies of excitation, and the procedure has shown exceptional sensitivity to defects which have internal free surfaces that are in close proximity to one another (such as delaminations).

Equipment and Experimental Technique

To apply mechanical excitation to a specimen for vibrothermography, a variety of equipment can be used. Any type of fatigue machine can be used to apply the lower frequency/higher amplitude type tests. The literature has reported the use of tension/compression axial machines and rotating-bar machines, for example. The higher frequency/lower amplitude tests can be performed on a variety of mechanical shakers driven by electromagnetic coils or piezoelectric transducers. The latter can be obtained to cover a wide range of excitation frequencies.

Once the specimen surface has been prepared to be nonreflective and to have as uniform a surface emissivity as possible (as described previously in the section on Thermography), the specimen is mounted in the mechanical exciter. For mounting in fatigue machines, standard specimen-mounting procedures should be followed. For the higher frequency/lower amplitude testing, a few simple procedures will help the first-time user to get started. An attachment jig is necessary that will both grip the specimen and allow it to be attached at one point to the shaker.[10-12] To improve the transmission of mechanical energy into the specimen, we have found it useful to apply an ultrasonic coupling agent between the specimen and the faces of the grips. By attaching the specimen at one point only, it is loaded inertially by the

---

Figure 9—Thermograph showing thermal pattern developed by a graphite-epoxy plate in global (plate) vibration mode

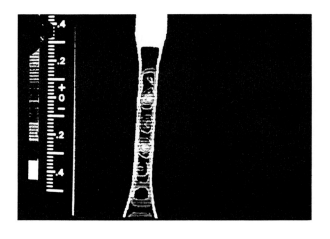

---

mechanical shaker. The amplitude of introduced strains is quite low using this method. The shaker can be driven by a frequency oscillator that is capable of slowly sweeping through a frequency range that is consistent with the band width of the shaker (Fig. 5). Once the sweep is begun, the operator observes the thermographic monitor and watches for thermal gradients to appear on the specimen. After completing the frequency sweep and noting those frequencies which excited thermal gradients and showed localized heating, the investigator can return to these particular frequencies, and more slowly investigate the development of the temperature patterns. One will find that the development of localized heating in this technique is very sensitive to the applied frequency so that, typically within ± 0.5 kHz of the resonant frequency, a temperature gradient will appear. At the resonant frequency, a maximum temperature is achieved as the frequency is swept past the value of frequency which excites the local flawed region. The temperature gradient disappears as the frequency changes away from resonance, again, usually greater than ± 0.5 kHz of the resonant frequency.

Various methods have been used to record the thermographic data from the monitor. Some video-thermographic systems have digital or analog magnetic recorders which can capture the thermographic pattern directly from the control monitor. We have been successful with recording the monitor screen using either a 35 mm camera or a video-television camera to take pictures from the screen.

References

1. Vanzetti, R., Practical Applications of Infrared Techniques, John Wiley and Sons, New York, 1972.

2. Wolfe, W., and Zissis, G.J., eds., The Infrared Handbook, The Infrared Information and Analysis Ctr., Environmental Research Inst. of Michigan, 1978

3. Bergstrom, L. and Borg, S.B., "Thermography in Real Time — Its Application to Non-Destructive Testing," Brit. J. Non-Destructive Testing, 10 (2).

4. Henneke, E.G. II, and Jones, T.S., "Detection of Damage in Composite Materials by Vibrothermography," Nondestructive Evaluation and Flaw Criticality for Composite Materials, ASTM STP 696 ASTM, Philadelphia, PA (1979).

5. Whitcomb, J.D., "Thermographic Measurement of Fatigue Damage," Composite Materials: Testing and Design, S.W. Tsai, ed., ASTM STP 674, ASTM, Philadelphia, PA 502-516 (1979).

6. Engelhardt, R.E. and Hewgley, W.A., "Thermal and Infrared Testing," Nondestructive Testing—A Survey, NASA SP-5113, U.S. Gov. Printing Office, 119-140 (1973).

7. Stalnaker, D.O. and Stinchcomb, W.W., "Load History-Edge Damage Studies in Two Quasi-Isotropic Graphite Epoxy Laminates," Composite Materials: Testing and Design (Fifth Conf.), S.W. Tsai, ed., ASTM STP 674, ASTM, Philadelphia, PA, 620-641 (1979).

8. Henneke, E.G., II, Reifsnider, K.L. and Stinchcomb, W.W., "Thermography—An NDI Method for Damage Detection," J. Metals, 31, 11-15 (Sept. 1979).

9. Reifsnider, K.L., Henneke, E.G., II, and Stinchcomb, W.W., "The Mechanics of Vibrothermography," The Mechanics of Nondestructive Testing, W.W. Stinchcomb, ed., Plenum Press, New York, 249-276 (1980).

10. Pye, C.J. and Adams, R.D., "Detection of Damage in Fibre Reinforced Plastics Using Thermal Fields Generated During Resonant Vibration," NDT Int., 14, 927-941 (1981).

11. Russell, S.S., "An Investigation of the Excitation Frequency Dependent Behavior of Fiber-Reinforced Epoxy Composites During Vibrothermographic Inspection," Ph.D. dissertation, Virginia Polytechnic Inst. and State Univ., Blacksburg, VA (1982).

12. Russell, S.S. and Henneke, E.G., II, "Dynamic Effects During Vibrothermographic NDE of Composites," NDT Int., 17, 19-25 (Feb. 1984).

13. Lin, S.S., "Frequency Dependent Heat Generation During Vibrothermographic Testing of Composite Materials," Ph.D. dissertation, Virginia Polytechnic Inst. and State Univ., Blacksburg, VA (1987).

14. McLaughlin, P.V., McAssey, E.V., and Dietrich, R.C., "Nondestructive Examination of Fibre Composite Structures by Thermal Field Techniques," NDT Int., 13, 56-62 (1980).

15. Wilson, D.W., and Charles, J.A., "Thermographic Detection of Adhesive-Bond and Interlaminar Flaws in Composites," EXPERIMENTAL MECHANICS, 276-280 (1981).

16. Reynolds, W.N. and Wells, G., "Video Thermography of Composite Materials," 4th Ann. Mtg. and 2nd Int. Mtg. on Comp. Mat., Napoli, April, 20-22, 1983.

Section VIIB

Adiabatic Thermoelastic Measurements

by C.E. Bakis* and K.L. Reifsnider**

Introduction

The adiabatic thermoelastic effect in elastic solids is the small, reversible temperature change resulting from the dilatational deformation of matter. If cyclic extensional strains are applied at a rate that is high enough to preclude heat transfer between elements of the material and their surroundings (i.e., ensuring adiabatic conditions), the temperature of the material will vary with the same wave form and frequency as the applied load. Pointwise temperature changes can then be related to stress and strain with the theory of thermoelasticity. Modern infrared radiometers have the sensitivity and response time required to measure these temperature changes on the surface of a stressed material. One such apparatus has been marketed by Ometron*** under the trade name SPATE (Stress Pattern Analysis by Thermal Emission) specifically for full-field stress measurements. The SPATE technique has been used successfully with homogeneous and heterogeneous materials to quickly evaluate stress and strain fields in a non-contact manner[1], and will be referred to presently to describe the measurement of adiabatic thermoelastic temperature changes in composite materials.

Theoretical Principles

Isotropic and Anisotropic Adiabatic Thermoelasticity

The adiabatic thermoelastic effect was first explained by Thomson (later to become Lord Kelvin) in 1853[2]. Using fundamental laws of thermodynamics, he derived expression (1), which relates the pointwise change in infinitesimal extensional strain components, ε^i $(i=1,2,3)$, in homogeneous, isotropic, linear-elastic matter and the small, reversible, adiabatic temperature change, Θ,

$$\Theta = - \frac{3T_o\alpha K}{c_\sigma} (\varepsilon_1 + \varepsilon_2 + \varepsilon_3) \qquad (1)$$

where T_o is the initial temperature of the material, α is the linear thermal expansion coefficient, K is the bulk modulus, and c_σ is the volumetric specific heat at

constant stress. The volumetric specific heat is found by multiplying the mass specific heat, c', by the mass density, ρ $(c = \rho c')$. Equation (1) can be written in terms of the change in extensional stress components, σ_i, as in Eq. (2).

$$\Theta = - \frac{T_o\alpha}{c_\sigma} (\sigma_1 + \sigma_2 + \sigma_3) \qquad (2)$$

Two well-known implications of Eqs. (1) and (2) are: (a) the temperature of isotropic matter with a positive thermal expansion coefficient increases with a negative dilatation, and decreases with a positive dilatation; and (b) a state of pure shear strain or stress produces no adiabatic thermoelastic temperature change in isotropic matter[3,4,5]. In situations where adiabatic conditions are not maintained or the elastic limit of the material is exceeded[6,7], the temperature variation is not reversible, and additional terms must be included in Eqs. (1) and (2) to account for heat transfer.

Biot gives the counterpart of Eq. (1) for anisotropic solids as

$$\Theta = - \frac{T_o}{c_\epsilon} \sum_{j=1}^{3} \sum_{j=1}^{3} \sum_{k=1}^{3} \sum_{l=1}^{3} \alpha_{kl} C_{ijkl} \varepsilon_{ij} \qquad (3)$$

where α_{kl} is the linear thermal expansion tensor, c_ϵ is the volumetric specific heat at constant strain, C_{ijkl} is the stiffness tensor, and ε_{ij} is the linear strain tensor[8]. Rewriting Eq. (3) in terms of the stress tensor yields Eq. (4).

$$\Theta = - \frac{T_o}{c_\epsilon} \sum_{k=1}^{3} \sum_{l=1}^{3} \alpha_{kl} \sigma_{kl} \qquad (4)$$

*Assistant Professor, Engineering Science and Mechanics, Pennsylvania State University, University Park, PA 16802.
**Reynolds Metals Professor, Engineering Science and Mechanics, Virginia Polytechnic Institute and State University, Blacksburg, VA 24061.
***Ometron Inc., 380 Herndon Pkwy., Suite 300, Herndon, VA 22070.

139

The difference between c_σ and c_ϵ for anisotropic solids, though quite small, is given by Eq. (5).[9]

$$c_\sigma - c_\epsilon = T_o \sum_{i=1}^{3} \sum_{j=1}^{3} \sum_{k=1}^{3} \sum_{l=1}^{3} \alpha_{ij}\alpha_{kl}C_{ijkl} \qquad (5)$$

Equations (3) and (4) imply that a state of pure shear strain or stress in anisotropic matter can result in a non-zero adiabatic thermoelastic temperature change if there exists a non-zero shear-extension coupling term in the thermal expansion tensor or the stiffness tensor. Equations (1) - (4) can be recast in the form given by Eq. (6) in order to express the adiabatic thermoelastic temperature change in terms of the two planar components of extensional stress acting on the surface of a solid.

$$\Theta = K_1\sigma_1 + K_2\sigma_2 \qquad (6)$$

Here, K_1 and K_2 represent the influence of the thermoelastic constants in the orthogonal 1 and 2 directions, respectively, for a particular initial temperature, T_o. The 1 and 2 directions correspond to the principal material directions in orthotropic matter (such as unidirectionally-reinforced composite materials). If the material is isotropic, $K_1 = K_2$, and Θ is proportional to the sum of the extensional stresses; otherwise, $K_1 \neq K_2$ and Θ is proportional to a nonuniformly-weighted sum of the extensional stresses. An apparent limitation of the adiabatic thermoelastic measurement technique is that the two stress components in Eq. (6) cannot be individually calculated from a given temperature change except in special cases where one component is known by a boundary condition[10]. For a series of measured temperature changes, Θ_i, the locus of possible combinations of σ_1 and σ_2 can be graphically represented by one of a series of parallel lines in the $\sigma_1 - \sigma_2$ plane (Fig. 1). The line passing through the origin suggests that in addition to the null stress state there are an infinite number of stress states resulting in no adiabatic thermoelastic temperature change.

Effect of Heterogeneity, Lamination, and Stacking Sequence

Equation (6) can be used to describe the "smeared" thermoelastic response of a composite material if the effective (average) thermoelastic constants of the material are known. In those situations where the constitution or relative amount of each phase of the composite is variable, a more general approach is to evaluate the temperature change in each constituent separately using Eq. (3) or (4), and to combine these changes in some manner to arrive at the net temperature change of the composite. Expression (7) represents one method of computing a weighted average of the several temperature changes in a non-layered composite,

$$\Theta^{net} = \sum_{j=1}^{n} \Theta^j X^j \qquad (7)$$

where Θ^j is the temperature change of the j-th constituent, X^j is an "influence factor" for the j-th constituent (such as the volume fraction), and n is the total number of constituents in the composite. A difficulty associated with using Eq. (7) is that an accurate micromechanical model for constituent extensional strains or stresses is required to compute each Θ^j.

Fig. 1—Graphical representation of the locus of possible values of extensional stress components for a measured temperature change

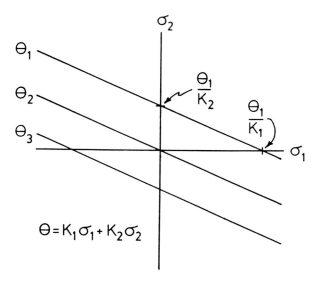

Laminated composite materials consist of multiple orthotropic layers arranged to meet specific strength and stiffness requirements. Considering that photons from a depth of only a few microns (μm) reach the infrared detector[11], and that there is no heat transfer between layers of a laminate during an adiabatic deformation, it is obvious that SPATE measurements are most sensitive to the deformation of the top ply on the surface of observation (assuming its thickness is greater than a few microns). The surface layer of matrix material commonly found on polymer composites can also influence thermal emissions, depending on the thickness of the layer. Since the deformations and resulting temperature changes of the constituent phases of the composite will differ in layers of dissimilar orientation, laminate stacking sequence must be known to interpret the measurements. That is, Eq. (7) needs to be evaluated only for the surface ply. Of course, the deformation of the surface ply reflects the deformation of the entire laminate if the plies remain bonded together during the adiabatic thermal emission process. Recently, Lesniak[12] has demonstrated that stresses a short distance below the surface of a homogeneous aluminum bar in four point bending can be measured with the SPATE technique if the heat transfer mechanisms are correctly modeled, but no analogous analysis for composite laminates is known to the authors, at this time.

Fig. 2—Theoretical effect of ply orientation on the adiabatic thermoelastic temperature change in unidirectional graphite/epoxy under constant axial stress or strain

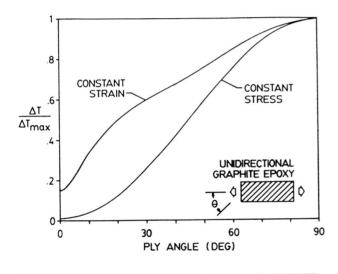

Fig. 3—Theoretical effect of ply orientation on the adiabatic thermoelastic temperature change in angle-ply graphite/epoxy laminates under constant axial stress or strain

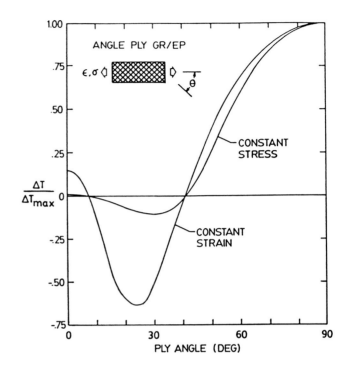

Fig. 4—Theoretical effect of surface ply orientation on the surface adiabatic thermoelastic temperature change in a quasi-isotropic graphite/epoxy laminate

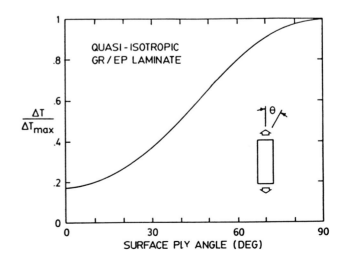

To illustrate the effect of surface ply orientation on the thermoelastic temperature change measured via infrared thermography, several theoretical calculations for continuous fiber graphite/epoxy laminates are presented next. (Experimental data were not available, and could differ somewhat from the present predictions which were calculated using the approach of Eq. 7). The material properties of graphite/epoxy were used in the calculatons because of their highly anisotropic nature[13,14]. Figure 2 is a prediction of the effect of ply orientation on the adiabatic temperature change for a unidirectional ply under a constant axial stress or strain change. There is a monotonic increase in temperature change as the ply orientation, θ, is changed from 0-deg. to 90-deg. Analogous behavior is predicted for $\pm\theta$ angle-ply laminates in Fig. 3. Note that there are two values of θ resulting in no thermal emission for angle-ply laminates, according to the calculations. In Figs. 2 and 3, the global transverse strains of the laminate varied with θ because of the dependence of the global Poisson's ratio on θ. In order to separate the effects of surface ply orientation and global Poisson's ratio, the surface ply angle in a quasi-isotropic laminate was varied while maintaining a constant global strain field (Fig. 4). Again, the temperature change is greatest when the surface ply orientation is perpendicular to the load direction. As a final example, the effect of laminate Poisson's ratio on the temperature change in a 0-deg. surface ply under a constant longitudinal strain change is examined (Fig. 5). The temperature change of the surface ply can be either greater than zero, less than zero, or equal to zero, depending on the magnitude of the transverse strain. This effect is caused by the opposite signs of a graphite/epoxy ply's thermal expansion coefficients in the longitudinal and transverse directions (typically -0.8 μ/K and 24 μ/k, respectively.)

Fig. 5—Theoretical effect of laminate Poisson's ratio on the adiabatic thermoelastic temperature change in a 0-deg graphite/epoxy ply

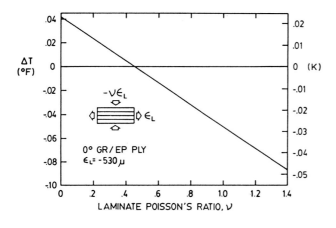

Infrared Radiometry

Although the feasibility of using high sensitivity, low response time thermocouples to measure the thermoelastic effect in solids has been demonstrated[7,15], there are significant advantages in using scanning infrared radiometry. In addition to being full-field and non-contact, the technique is increasingly sensitive to stress as the temperature of the observed material increases — a feature that is attractive for high temperature applications. Indeed, the adiabatic temperature change for a given stress change increases linearly with T_o, as in Eqs. (1)-(4), and the attendant change in photon emittance detected by the radiometer, δQ (uncorrected for detector response), increases with the square of T_o, as in Eq. (8),

$$(\delta Q) = 3\,e\,B\,T_o^2\,\Theta \tag{8}$$

where e is the surface emissivity and B is the Stefan-Boltzmann constant for infrared photon radiation[11]. Therefore, the overall stress sensitivity of the SPATE apparatus increases with the cube of the material's temperature, as in Eq. (9).

$$\text{SPATE signal} \propto T_o^3\,(K_1\sigma_1 + K_2\sigma_2) \tag{9}$$

Reference 1 has some examples and limitations of the technique at very high temperatures.

Experimental Apparatus and Procedure

Thermographic Equipment

The apparatus used to measure the small temperature changes, an Ometron SPATE 8000, consists of a scanning infrared photon detector coupled to a correlator (lock-in amplifier) and computer. The sensitivity of the system is 0.001 K — approximately equivalent to the uniaxial stresses for various materials shown in Fig. 6. The functions of the computer and correlator are to control the camera scan activities and condition the measured photo emittance such that the sinusoidal temperature variation occurring at the same frequency as the sinusoidal load can be determined. The correlator rejects a temperature change with no sinusoidal content at the test frequency, such as a quasi-steady state temperature increase caused by dissipative heating. The camera scans the test specimen pointwise in a raster-like manner, enabling the computer to store the recorded signal at each point as a digital quantity. The smallest area that can be sampled at each point in the scan area is a .5-mm-diameter circle. If the raster scan takes a long time to complete (up to two hours is common), it may be necessary to cycle the specimen at a load that is low enough to inhibit damage development during the scan. Two factors influencing the scan time are the scan resolution (the number of data points to be recorded over a given area) and the sample time (the amount of time spent acquiring data at a single position on the specimen). In situations where the measured signal is small, electronic averaging must be applied over a selectable time period associated with a filter "time constant" to reduce high frequency noise. Averaging is necessary since only a single value of temperature change is recorded at each point. As the time constant increases, the amount of time spent sampling data at a particular point on the specimen must also increase. Once a scan has been completed, the digital information may be stored on a magnetic disk for future exfuture examination. A video monitor enables the operator to observe the results of the scan as each point in the scan is sequentially displayed on a color-coded contour map of temperature change.

Experimental Procedure

A variety of material systems, including several graphite/epoxy systems, graphite/PPS, aramid/epoxy, boron/aluminum, and glass/epoxy, to name a few, have been studied by the authors with the SPATE apparatus. Of these, fiber-dominated graphite/epoxy and aramid/epoxy are the most difficult to examine because of their relatively low temperature change during cyclic loading (Fig. 6). A detailed review of the procedure for obtaining high quality, repeatable data with fiber reinforced composites follows. Aspects of the procedure not detailed in the SPATE operator's manual will be emphasized presently. The manual should be consulted for routine procedures.

Fig. 6—Theoretical uniaxial stress resulting in an adiabatic thermoelastic temperature change of .001 K in several materials

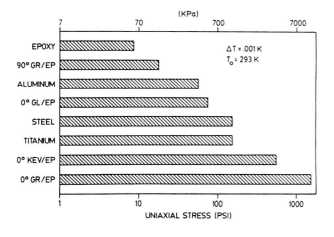

Mechanical Excitation

Any source of harmonic mechanical excitation can be used with the SPATE apparatus, although the high cyclic load amplitudes required to obtain a good temperature signal/noise ratio with fiber-dominated graphite/epoxy laminates are most readily obtained via a servo-hydraulic load frame equipped with a function generator. Because the SPATE lock-in amplifier output is proportional to the sinewave content of the temperature change, it is essential to maintain a constant load form for quantitative experiments. Unless specifics of the test dictate otherwise, one should maintain a load that does not cause

damage growth during the scan. It is also important to maintain loads within the linear-elastic range of the material if stresses or strains are to be computed with the linear theory reviewed earlier; otherwise, modifications to the theory must be made to account for the inelastic deformation[7]. In most instances, a sinusoidal load with a maximum of 30-40% of the specimen's tensile strength and a load ratio (min/max) of 0.1 is appropriate.

Fig. 7—Effect of loading frequency on the measured signal (temperature change) in several AS4/3502 graphite/epoxy lamiantes under constant load amplitude

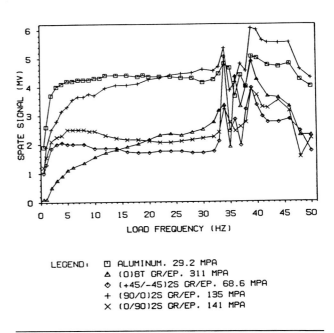

To ensure adiabatic deformations and the validity of Eqs. (1) - (4), it is necessary to increase the load frequency until no further increase in the SPATE signal is realized. Highly variable thermal emissions from several AS4/3502 graphite/epoxy laminates over a range of loading frequencies have been observed by the authors (Fig. 7), but this effect may be peculiar to the specific material interrogated. For example, Potter has noted a constant thermal emission for frequencies above 5 Hz with XAS/914 graphite/epoxy[16]. The wide variation in SPATE signal between 30 and 35 Hz in Fig. 7 is attributed to a resonance in the servo-hydraulic load frame. It is therefore desirable to maintain a constant loading frequency in quantitative comparisons of data. For the composite material systems studied by the authors, frequencies between 5 and 15 Hz are adequate. Higher frequencies sometimes cause phase shifts in the cyclic temperature as a function of position on coupon-type specimens, resulting in a slight artificial gradient in the "X" and "Y" signal outputs along the length of the specimen. At this time, the cause of the phase shift remains undetermined, but it may be related to the dynamic response of the load frame. Scanning in the "R" output mode of the lock-in amplifier ($R = \sqrt{X^2 + Y^2}$) bypasses the phase shift problem, but obscures the distinction between tensile and compressive stresses in test specimens with complex shapes.

Specimen Preparation:

A flat-finish paint can be applied to the specimen in order to obtain a diffuse surface and a uniformly high surface emissivity in the infrared spectrum. This procedure reduces the possibility of reflected heat sources being modulated at the test frequency, and maximizes the sensitivity of the measurements. Krylon[TM] ultra-flay spray paint applied in two thin coats, for example, serves this purpose well[1], and is easily removed with acetone. Where possible, it is wise to avoid the use of paint since investigators have observed an attenuation of the photon emittance due to the paint's thickness. This effect is especially evident at increased frequencies (> 20 Hz) and paint thicknesses (> 12 spray passes)[17,18]. Epoxy-matrix composites that have a diffuse surface finish do not require any surface preparation.

SPATE Setup:

A constant distance from the detector to the specimen should be maintained in order to minimize variatons in the attenuation of the infrared radiation over the distance between the detector and the specimen. It is also advantageous to maximize the spatial resolution (i.e., minimize the distance between the detector and specimen) in order to observe the effect of localized damage that typically occurs in composites prior to catastrophic failure. With a proper surface preparation, the angle of obliquity between the detector and the specimen can be as high as 55 deg. before measurement inaccuracies become significant[17].

When analyzing graphite/epoxy or aramid/epoxy composites, typical output voltage amplitudes from the lock-in amplifier range from 0 to ± 10 mV. (Some graphite/epoxy laminates have no measurable temperature change during cyclic loading). The low-pass filter time constant for this range of signal should be at least 0.1 to 0.3 sec. to produce an adequate signal/noise ratio and stable output. The sample time should then be set at 3 to 10 times the time constant for an accurate spatial resolution of temperature change on the color monitor. If the sample time is too short, there is not sufficient time for the signal output to stabilize at the new value at each subsequent sample point. Hence, sample times of 0.3 to 3 sec. are typical — leading to very long scan times with these "worst-case" material systems.

Adjustment of the electronic "zero" of the lock-in amplifier is essential for accurate stress analysis. After properly adjusting the phase of the lock-in amplifier with the reference signal from the function generator of the load frame, the zero can be set by performing line scans across several sections of the area of interest with no load applied to the specimen. In order to carry out such a no-load scan, though, the reference signal must still be supplied to the lock-in amplifier. If there is a variation of the SPATE signal along any section of the specimen during this procedure, the optimum adjustment is such that the *average* signal is zero.

During a scan, one can expect to obtain spurious data when the focal area lies partly on and partly off of an edge of the specimen. These data points are inaccurate because the stressed material cyclicly enters and leaves the focal area, resulting in a false apparent temperature change.

Calibration of the SPATE thermal emission scale can be accomplished with the use of high sensitivity, low response time thermocouples placed at several locations on the specimen. However, care must be exercised to ensure that all test variables mentioned previously are held constant for a particular calibration.

Interpretation of Results

It was mentioned earlier that it is the response of the surface ply that dominates infrared temperature measurements. Figure 8 illustrates this phenomenon in two center-notched quasi-isotropic, graphite/epoxy laminates with different stacking sequences: $(0,90,45,-45)_s$ and $(45,90,-45,0)_s$. The black color on the thermal emission scale corresponds to no temperature change during the load cycle. Colors above black quantify the amount of (uncalibrated) cooling during tensile loads, while colors below black quantify heating. In the first laminate, the disturbance in the temperature field caused by the stress concentration is symmetric about the hole, while in the second laminate the pattern is anti-symmetric. Despite the identical, symmetric global strains in these laminates, the anti-symmetric thermal emission pattern in Fig. 8b can be predicted with Eq. (3) or (4) by considering the anti-symmetric stress pattern about the hole in the 45-deg. surface ply (Fig. 9).

The utility of the SPATE technique for monitoring damage development in fiber composite laminates is exemplified by the penetrant-enhanced X-ray radiographs and SPATE thermographs in Figs. 10 and 11, respectively. The first example (Figs. 10a and 11a) involves the same $(0,90,45,-45)_s$ graphite/epoxy laminate as in Fig. 8a. The advanced damage condition around the notch, consisting of matrix cracks and delaminations, is characterized by a very low temperature change during the load cycle. Based on the information obtained in the radiograph, it is

Fig. 8—SPATE thermographs of two center-notched quasi-isotropic graphite/epoxy laminates with different stacking sequences: (a) (0,90,45, − 45),

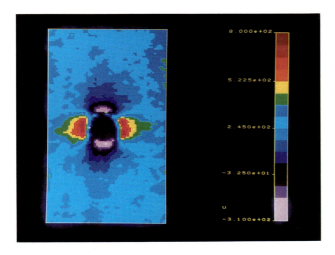

(b) (45,90, − 45,0),

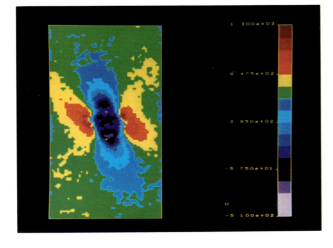

Fig. 8b. As in the previous example, damage in the surface ply consists of matrix cracks and delaminations. Each matrix crack on the surface of observation appears as a line of low thermal emission in the thermograph due to the relaxation of stress transverse to the cracks in the surface ply. Delaminated portions of the surface ply that are bounded on three sides by the hole boundary and two parallel matrix cracks in the surface ply carry no load and, therefore, have no thermal emission. Cox and Pettit[19] and Lohr[20,21] have also shown that the SPATE technique can be used to detect impact and fatigue damage in graphite fiber composite laminates.

Reduced thermal emissions in graphite/epoxy laminates can be attributed to a relaxation of some or all components of stress in the surface ply. Another cause for pointwise variations of the thermal emission in composites is the existence of manufacturing irregularities — particularly non-uniform phase and void distribution. For example, matrix-rich regions of a composite can be resolved from matrix-poor regions because of their different thermoelastic responses[22].

Fig. 9—Predicted contours of adiabatic thermoelastic temperature change in a center-notched (45,90, −45,0), graphite/epoxy laminate

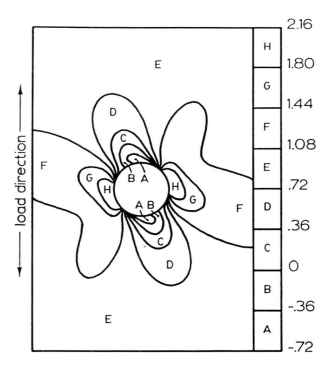

known that the 0-deg. surface ply is delaminated near the notch and has several associated cracks parallel to the fibers. It is therefore obvious that delaminated regions of this ply must be under a nearly-pure axial stress state because of the lack of transverse constraint from the adjoining sublaminate. The very low temperature change measured near the notch is thus caused not by the total absence of stress in the surface ply, but by the absence of transverse stress in that ply. The surface ply continues to bear load in the 0-deg. fiber direction, but since the coefficient of thermal expansion in this material system is extremely low in the fiber direction, there is little resultant thermal emission. The second example (Figs. 10b and 11b) involves the same (45,90,-45,0), laminate as in

145

Fig. 10—Penetrant-enhanced X-ray radiographs of center-notched graphite/epoxy laminates with extensive fatigue damage:

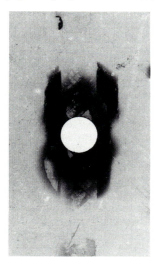 

(a) (0,90,45, −45)ₛ   (b) (45,90, −45,0)ₛ

Fig. 11—SPATE thermographs of center-notched graphite/epoxy laminates with extensive fatigue damage:

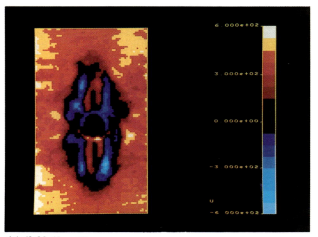

(a) (0,90,45, −45)ₛ

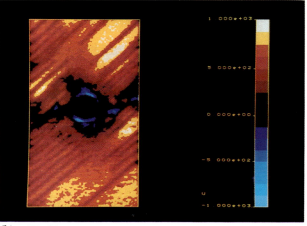

(b)   (45, 90, −45,0)ₛ

References

1. Oliver, D.E., "Stress Pattern Analysis by Thermal Emission," Ch. 14 in the SEM "Handbook on Experimental Mechanics", A.S. Kobayashi, Ed., Prentice-Hall (1987), pp. 610-620.

2. Thomson, W., "On the Dynamical Theory of Heat," Trans. Roy. Soc. Edinburgh, 20 (1853), pp. 261-288.

3. Joule, J.P., "On Some Thermo-dynamic Properties of Solids," Philos. Trans. Roy. Soc., 149 (1859), pp. 91-131.

4. Compton, K.T. and Webster, D.B., "Temperature Changes Accompanying the Adiabatic Compression of Steel," Phys. Rev., Series 2, 5 (1915), pp. 159-166.

5. Stanley, P. and Chan, W.K. "Quantitative Stress Analysis by Means of the Thermoelastic Effect," J. Strain Anal., 20 (1985), pp. 129-137.

6. Enke, N.F. and Sandor, B.I., "Cyclic Plasticity Analysis by Differential Infrared Thermography," Proc. VI Intl. Congr. Exper. Mech., Vol. II, Portland, OR, 6-10 June 1988, SEM, Bethel, CT, pp. 837-842.

7. Jordan, E.H. and Sandor, B.I., "Stress Analysis from Temperature Data," J. Testing and Eval., JTEVA, 6 (1978), pp. 325-331.

8. Biot, M.A., "Thermoelasticity and Irreversible Thermodynamics," J. Appl. Phys., 27 (1956), pp. 240-253.

9. Nowacki, W., Dynamic Problems of Thermoelasticity, Noordhoff (1975).

10. Huang, Y.M., Hamdi AbdelMohsen, M.H., Lohr, D., Feng, Z., Rowlands, R.E., and Stanley, P., "Determination of Individual Stress Components from SPATE Isopachics Only," Proc. VI Intl. Congr. Exper. Mech., Vol. I, Portland, OR, 6-10 June 1988, SEM, Bethel, CT, pp. 578-584.

11. Hudson, R.D., Jr., Infrared System Engineering, Wiley, New York (1969).

12. Lesniak, J.R., "Internal Stress Measurements," Proc. VI Intl. Congr. Exper. Mech., Vol. II, Portland OR, 6-10 June 1988, SEM, Bethel CT, pp. 825-829.

13. Chamis, C.C., "Simplified Composite Micromechanics Equations for Hygral, Thermal, and Mechanical Properties," SAMPE Qtly., 15 (1984), pp. 14-23.

14. Jones, R.M. Mechanics of Composite Materials, McGraw-Hill (1975).

15. Neubert, H., Schulte, K., and Harig, H., "Evaluation of the Damage Development in CFRP by Monitoring Load Induced Temperature Changes," Composite Materials: Testing and Design (9th Symp.), STP 1059, S.P. Garbo, Ed., ASTM, Philadelphia, in press.

16. Potter, R.T., "Stress Analysis in Laminated Fiber Composites by Thermoelastic Emission," Proc. 2nd Intl. Conf. on Stress Anal. by Thermoelastic Tech., Paper No. 20, 17-18 Feb. 1987, London.

17. Stanley, P. and Chan, W.K., "SPATE Stress Studies of Plates and Rings Under In-Plane Loading," Exp. Mech., 26 (1986), pp. 360-370.

18. Belgen, M.H., "Structural Stress Measurements with an Infrared Radiometer," ISA Trans., 6 (1967), pp. 49-53.

19. Cox, B.N. and Pettit, D.E., "Nondestructive Evaluation of Composite Materials Using the Stress Pattern Analysis by Thermal Emissions Technique," Proc. SEM Spring Conf., Houston, TX, 14-19 June 1987.

20. Lohr, D.T. and Sandor, B.I., "Impact Damage Analysis by Differential Infrared Thermography," Proc. SEM Fall Conf., Savannah, GA, 1987.

21. Lohr, D.T., Enke, N.F., and Sandor, B.I., "Analysis of Fatigue Damage Evolution by Differential Infrared Thermography," Proc. SEM Fall Conf., Savannah, GA, 1987.

22. Bakis, C.E. and Reifsnider, K.L., "Nondestructive Evaluation of Fiber Composite Laminates by Thermoelastic Emission," Review of Progress in Quantitative NDE, 7B, D.O. Thompson and D.E. Chimenti, Eds., Plenum, 1988, pp. 1109-1116.

Section VIIC

Damage Evaluation by Laminate Deply

by Charles E. Harris

Introduction

The specimen-deply technique is a simple destructive-examination method whereby the individual plies of a laminate are separated for easy examination. This technique is excellent for determining the laminate-stacking sequence, studying the details of broken fibers, and for determining the extent and precise interface location of delaminations. The interlaminar-bond strength is broken down by a partial pyrolysis of the matrix resin between the plies of the laminate. This allows for the separation (unstacking) of the individual plies provided the adjacent plies have different fiber angles. The individual plies may then be microscopically examined to study the details of broken fibers. Furthermore, by using a marking agent prior to the pyrolysis, delaminations are clearly visible and matrix cracks in the adjacent ply may also be visible. This procedure is well suited to general angle-ply laminates of graphite/epoxy. The author's procedure for graphite/epoxy described below was originally developed by S.M. Freeman[1].

Specimen Preparation and Deply

The only specimen preparation required prior to the pyrolysis process is to expose the damaged regions to the marking agent. A 9.2-weight-percent solution of gold chloride in diethyl ether with a few drops of photo-flow 200 has been found to be an effective marking agent. The gold-chloride solution is applied to the specimen in a manner similar to the use of zinc iodide for X-ray radiography. For example, if the damage is associated with a notch such as a circular cutout, a reservoir for the gold-chloride solution can be formed by covering one surface of the specimen at the hole. After subjecting the specimen to the gold-chloride solution for about 30 min., the remaining solution is collected for recycling and the excess diethyl ether is driven off to prevent gas-bubble formation during the pyrolysis. This may be accomplished by heating the specimen to 120-140°F for about an hour.

The best results for graphite/epoxy, T300/5208 and AS4/3502, have been obtained at a pyrolysis temperature of approximately 785 ± 25°F (418 ± 15° C). The time required for an eight-ply specimen is approximately 30 min. and the time required for thicker laminates may be estimated by multiplying 30

min. by the ratio of the number of plies to eight. The author has used an electric tube-style oven with an argon-gas environment for the pyrolysis. This setup is shown in Fig. 1. The chamber is continuously vented to the outside. Chamber venting is essential to remove the noxious fumes emitted by the specimen during the pyrolysis process. An uncontrolled cool down is accomplished by simply turning off the oven and leaving the specimen in the oven until it can be comfortably removed.

Fig. 1—Tube-style oven with chamber purging arrangement

The specimen is supported in the oven by the boat-style holder shown in Fig. 2(a). The specimen rests on a heat-treated steel-wire mesh as shown in Fig. 2(b) which shows the holder being inserted into the oven tube. The holder helps to collect the carbonized residue from the pyrolysis and provides a convenient method for handling the delicate specimen after the pyrolysis process is complete.

Fig. 2(a)—Specimen holder

Fig. 2(b)—Specimen holder with specimen being inserted into oven tube

The pyrolyzed specimen is quite delicate and must be handled with care. This is especially true of plies with fiber angles greater than 15-20 deg. Thicker specimens, as illustrated in the before and after views in Fig. 3, typically have a little more adhesion between the plies than do thinner specimens which can almost disintegrate if mishandled. Therefore, it is recommended that adhesive tape be used to reinforce each ply during separation to avoid damaging the plies during deplying. For example, a layer of tape may be applied to the outside surface of the first ply. The first ply is then peeled away from the laminate

exposing the interface between the first two plies. A layer of tape would then be applied to the exposed surface of the second ply which is peeled from the laminate exposing the interface between the second and third ply. Proceeding in this manner, each ply is reinforced by tape on one surface and has an exposed surface which allows for individual ply and interface damage evaluation. As a final observation about unstacking the plies, there may be local regions of adhesion between adjacent plies which must be broken. This can be accomplished by a little prodding using a thin-bladed exacto knife.

Fig. 3—Cross-sectional view of specimen before and after pyrolysis

Damage Evaluation

A comprehensive evaluation of damage typically requires a combination of low magnification (20-30 X) and high magnification (200-400 X). Examinations may be accomplished by an optical microscope or a scanning-electron microscope (SEM). When using an optical microscope, illumination of the specimen by the proper angle of reflected light is essential. The gold-chloride particles reflect the light differently than the specimen background. This is also true of the discontinuities formed by broken fibers. Figure 4(a) shows a typical delamination region viewed in an optical microscope at 20 X magnification. This delamination at the 0/+45 interface of a [0/±45/90]$_s$ laminate with a machined crack-like notch is relatively large, so the gold chloride particles are quite dense. Figure 4(b) shows a line of broken fibers in the 0-deg ply of a notched [0/90]$_s$ laminate viewed at 20 X in an optical microscope. While the line of broken fibers is clearly visible, the details of individual broken fibers cannot be studied at this low magnification.

Optical microscopes are sufficiently powerful to study individual fiber fracture. However, a difficulty arises because the complete field of vision is usually not in focus because of the natural contours of the ply surface. This problem is avoided by using the SEM. Typical SEM photographs showing the details of broken fibers are shown in Figs. 5(a) and 5(b) at 30

Fig. 4—Typical optical photographs of laminate deply results as viewed in anoptical microscope at 20X

a. Notch-tip delamination at a 0/+45 interface in a [0/+45/90]ₛ laminate

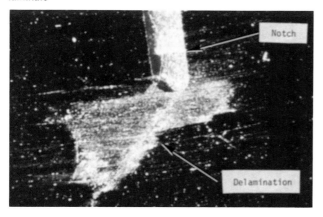

b. Line of broken fibers extending from the notch in a 0° ply of a [0/90]s

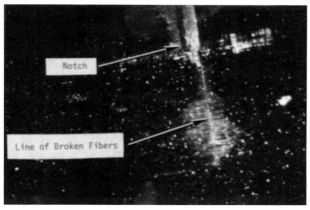

X and 200 X magnification. At 200 X magnification, individual fibers are clearly visible. These images were enhanced by coating the viewing surface with a thin (≅300 A) coating of gold which increases the number of free electrons released when the SEM voltage is applied to the specimen. Using this procedure, the photographs in Figs. 5(a) and 5(b) are views of the surface topography. As an alternative, noncoated specimens may be examined by the SEM to provide a spectrum analysis of the chemical composition of the surface. This procedure will identify the location and density of the gold-chloride particles from the marking agent.

Conclusions

The laminate-deply technique is a powerful yet simple to use method of destructively examining a specimen to determine the precise damage state. The method is especially useful in determining the extent and interface location of delaminations and for the study of fiber fracture. The method has been used extensively for the investigation of damage in continuous fiber graphite/epoxy angle-ply laminates.

Fig. 5—Typical photographs of deply results as viewed in a scanning-electron microscope

(a) Line of broken fibers extending from the notch of a 0-deg ply of a [0/+45/90], laminate at 30 X

(b) Line of broken fibers extending from the notch in a 0-deg ply of a [0/90]₂ₛ laminate at 200 X

References

1. Freeman, S.M., "Characterization of Lamina and Interlaminar Damage in Graphite-Epoxy Composites by the Deply Technique," Composite Materials: Testing and Design (Sixth Conf.), ASTM STP 787, ed. I.M. Daniel, ASTM, 50-62 (1982).

Section VIIE

Vibration-Test Methods for Dynamic-Mechanical-Property Characterization

by Ronald F. Gibson*

Introduction

Dynamic-mechanical properties (elastic modulus and internal damping) of materials may be characterized using either vibration or wave-propagation experiments. This section deals only with vibration-test methods, which cover the nominal frequency range 0.001-1000 Hz. Wave-propagation methods are not discussed here.

Measured dynamic-mechanical properties of structural materials are used not only as input data for design equations, but as quality-control parameters during fabrication and as nondestructive-evaluation parameters during in-service inspections. Valid dynamic-mechanical-property measurements are not easy to obtain, even on conventional structural materials; the unique properties of composite materials make such measurements even more difficult. Experimental approaches range from laboratory bench-top methods to portable field-inspection techniques.

This chapter describes the most commonly used techniques and discusses the special problems associated with each method. Several previous survey articles and a book dealing with dynamic mechanical behavior of composites may also be of interest to the reader[1-4].

Although standards have been developed by the American Society for Testing and Materials (ASTM) for measurement of dynamic-mechanical properties of low modulus polymers[5] and add-on damping treatments consisting of high-damping polymers[6], none exists specifically for composite materials. Problems encountered in applying some of these standard test methods to composites are also discussed.

Complex Modulus Notation

While the assumption of linear-elastic behavior is normally the basis for static-mechanical-property testing, the assumption of linear-viscoelastic behavior is usually the basis for dynamic-mechanical-property testing. Polymer-matrix composites in particular are known to exhibit viscoelastic behavior, which causes energy dissipation (damping) and frequency dependence of both stiffness and damping during dynamic testing. Such behavior is most often observed in matrix-dominated response under shear or off-axis loading. The assumption of linearity of dynamic-viscoelastic response is valid when both stiffness and damping are independent of vibration amplitude.

Complex modulus notation is often thought of as just a mathematically convenient way of combining stiffness and damping in one expression, but it does have a basis in viscoelasticity theory. The most general stress-strain relationships for a linear-viscoelastic anisotropic material may be represented by the well-known hereditary-integral formulation of the Boltzmann superposition principle[7,8]. While such equations are useful for describing creep or relaxation during static testing, a different form is more useful for dynamic testing. Using a contracted subscript notation[9] and the assumption that stresses and strains vary sinusoidally with time, the heriditary-integral formulation reduces to:

$$\tilde{\sigma}_p(t) = C_{pq}^*(\omega)\tilde{\varepsilon}_q(t) \qquad (1)$$

where

$$
\begin{aligned}
p,q &= 1,2,\ldots 6 \\
\tilde{\sigma}_p(t) &= \text{sinusoidally varying stress components} \\
\tilde{\varepsilon}_q(t) &= \text{sinusoidally varying strain components} \\
\omega &= \text{frequency} \\
t &= \text{time} \\
C_{pq}^*(\omega) &= \text{complex modulus}
\end{aligned}
$$

One of the results of this development is that the frequency-domain complex modulus is related to the time-domain relaxation modulus through a Fourier transform[7,8].

The complex modulus can be expressed as

$$C_{pq}^*(\omega) = C'_{pq}(\omega) + i\,C''_{pq}(\omega) = $$
$$C'_{pq}(\omega)\,[1 + i\eta_{pq}(\omega)] \qquad (2)$$

*Mechanical Engineering Department, University of Idaho, Moscow, Idaho. This section was written while the author was on sabbatical leave at the Composite Materials and Structures Center, Michigan State University, East Lansing, Michigan.

where

$C'_{pq}(\omega)$ = storage modulus
$C''_{pq}(\omega)$ = loss modulus
$\eta_{pq}(\omega)$ = loss factor = $C''_{pq}(\omega)/C'_{pq}(\psi)$ = tan $\delta_{pq}(\omega)$
$\delta_{pq}(\omega)$ = phase angle between $\tilde{\sigma}_p(t)$ and $\tilde{\varepsilon}_q(t)$
i = imaginary operator = $\sqrt{-1}$

The storage modulus and the loss factor are actually measured during dynamic-mechanical testing, whereas the loss modulus is a derived property. Although the frequency dependence of the complex modulus is assured by the mathematical development, the complex moduli of polymeric materials are also known to depend on environmental conditions such as temperature and moisture.

The fact that eq. (1) has the same form as a linear-elastic stress-strain law has led to the development of a correspondence principle.[10,11] Using this principle, the corresponding viscoelastic forms of other linear-elastic constitutive relationships (e.g., those for the orthotropic lamina or the general laminate) may be found. For example, the dynamic behavior of the orthotropic lamina can be characterized by such properties as the complex-longitudinal modulus, $E_1^*(\omega)$, and the complex-transverse modulus, $E_2^*(\omega)$, while laminate behavior can be characterized by such properties as the complex-extensional stiffness, $A_{ij}^*(\omega)$, and the complex-flexural stiffness, $D_{ij}^*[\omega]$. Obviously, a wide variety of experiments with different stress states is necessary to characterize all of the complex moduli for an anisotropic composite material.[3]

Finally, although sinusoidally varying deformations were assumed in the development, it has been shown that as long as stiffness and damping show some frequency dependence, the complex-modulus notation is also valid for nonsinusoidally varying deformations.[12] Anomalous analytical results such as noncausal response can occur if the components of the complex modulus are independent of frequency. This turns out to be an academic problem, however, since polymeric materials do have frequency-dependent complex moduli.

Fig. 1—Comparison of loss-factor data from three different test methods with thermoelastic predictions for the same aluminum-cantilever-beam specimens. From Ref. 16

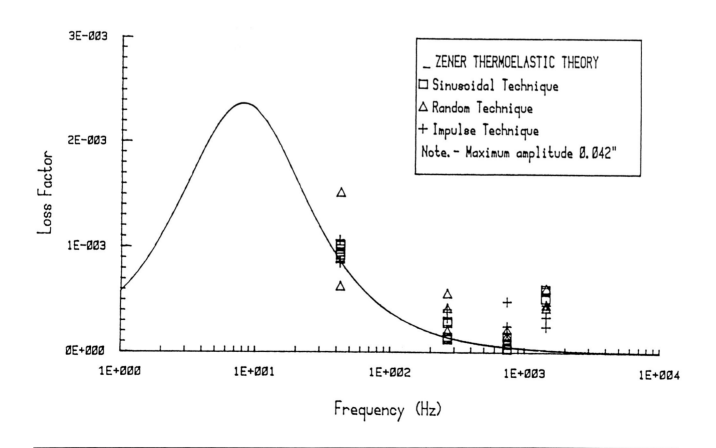

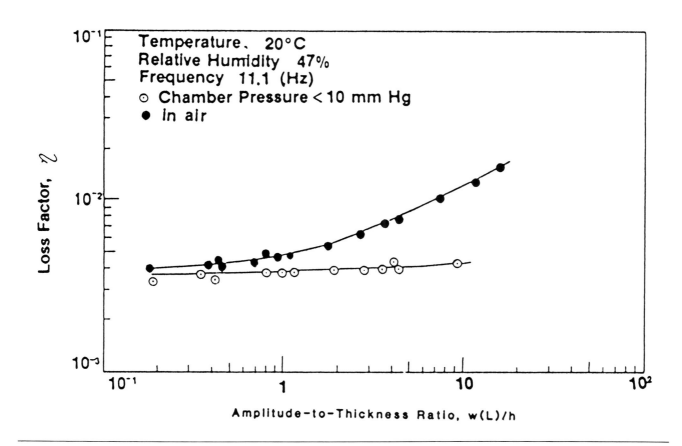

Fig. 2—Variation of loss factor with amplitude-to-thickness ratio for chopped E-glass/polyester composite in air and in vacuum. From Ref. 17

Special Considerations

Before getting into the details of the various test techniques, it is appropriate to discuss some of the problems that are likely to be encountered during the tests and subsequent data reduction. Some of the problems are inherent in dynamic-mechanical testing of any material (but are worthy of mention again here), and some are unique to composites.

Parasitic damping is a collective term for all of the extraneous energy dissipation that occurs during a dynamic-mechanical test. Common examples are air damping due to aerodynamic drag on the specimen, acoustic radiation, and friction damping at specimen support points and transducer attachments. Because of parasitic damping, the measured damping values will always be greater than the actual material damping. Great care must be taken to insure that parasitic damping has been reduced to an acceptable level before reporting damping data.

Fortunately, most of the parasitic-damping mechanisms are nonlinear (i.e., the damping depends on amplitude), whereas the viscoelastic damping in undamaged polymers due to relaxation and recovery of

the molecular network following deformation is linear. Thus, a check of amplitude dependence of damping can be used to detect parasitic damping. Aluminum or steel calibration specimens are often used to establish confidence in damping measurements because thermoelastic theory predicts the material damping quite accurately, and because the damping in such metals is much lower than that for composites.[13-18] Thermoelastic damping involves a different kind of relaxation mechanism (that of deformation-induced heat flow), which is also linear. Cross verification of damping measurements using several different techniques is highly recommended.[16,19] For example, Fig. 1 shows the results from calibration experiments with aluminum using three independent techniques, along with thermoelastic predictions.

Cantilever-beam specimens vibrating in flexure may be subjected to significant air damping. As shown in Fig. 2, the difference between damping in air and in vacuum increases with tip amplitude. These data provide a vivid example of the linearity of material damping and the nonlinearity of air damp-

153

ing. Thus, the tip amplitudes should be less than the beam thickness if the tests are to be conducted in air—otherwise, the test should be conducted in a vacuum.

Friction damping at specimen-support points can be minimized by attaching supports at nodal points for the vibrational mode of interest, or by using stress-relief shoulders on the specimen to shift the clamping surface away from the region of high stress. Noncontacting response transducers such as eddy current, electro-optical or capacitance probes can be used to eliminate damping due to motion of transducers and associated lead wires. The added mass of transducers such as accelerometers may have a significant effect on measured resonant frequencies and corresponding modulus calculations.

The stiffness of the test apparatus should be much greater than that of the specimen so that most of the deformation occurs in the specimen during the test; otherwise, both modulus and damping measurements will be invalid. For example, the commercially available dynamic-mechanical analyzers described in the ASTM standards[5] were developed for testing low-modulus polymers, and the stiffness of the specimen mounting hardware is generally insufficient for accurate determination of dynamic properties of high-modulus composites. In order to reduce the specimen stiffness to the range required for valid data with these devices, it may be necessary to use specimen thicknesses on the order of the single-ply thickness, so that testing of multi-ply laminates may not be possible. Since many of these devices operate in a flexural mode, laminates that produce coupling between bending and twisting and between bending and extension (e.g., unsymmetric layups) should be tested with caution; the equations used to convert measured specimen resonant frequencies to storage moduli are usually based on the assumption of pure bending.

A related problem is the transverse-shear effect in high-modulus composite specimens. The assumption of pure bending is part of the Bernoulli-Euler beam theory, whereas the effects of transverse shear and rotary inertia are included in Timoshenko beam theory[20-21]. Transverse-shear effects have been shown to be more significant for materials having high ratios of extensional modulus to through-the-thickness shear modulus, E/G; this ratio is at least ten times higher for high-modulus composites than for conventional metallic materials. Sandwich panels with honeycomb or foam cores have even higher E/G ratios due to the low shear modulus of the core material. Transverse-shear effects show up at high frequencies, which are generated by testing specimens in higher modes or shorter lengths. Figure 3 shows the storage modulus and loss factor for a wood specimen based on the use of Bernoulli-Euler theory and Timoshenko theory.[22] Note that the apparent modulus decreases and the apparent loss factor increases with frequency for the Bernoulli-Euler theory, but these changes are not real. In fact, both modulus and loss factor are essentially independent

of frequency, as shown by using Timoshenko theory in the data reduction. It appears that the length-to-thickness ratio, L/t, for highly anisotropic beam specimens must be at least 100 in order to minimize shear effects in lower modes.[23] Figure 4 shows the theoretical modulus correction factors for a cantilever beam having $L/T = 100$ vs. mode number for several E/G ratios.[23] Similar correction factors for the loss factor are given in Ref. 24. The use of high-aspect ratio specimens may also be necessary in order to minimize end effects. Although no work has been done on end effects in dynamic testing, analysis of an anisotropic strip under static end loading has shown that the decay length for end effects is much greater than that predicted by St. Venant's principle.[25-27]

Fig. 3—*Frequency dependence of storage modulus and loss tangent for wooden beams according to Bernoulli-Euler beam theory (●) and Timoshenko beam theory (O). From Ref. 22*

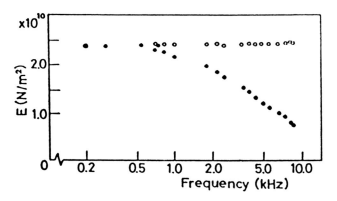

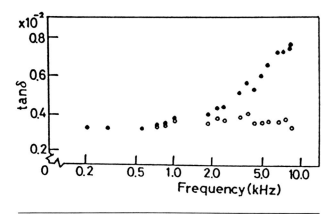

Finally, it should be remembered that the measured properties of the heterogeneous composite are only 'effective' properties of an equivalent homogeneous material. That is, an 'effective modulus' relates the volume averaged stress to the volume averaged strain over some representative

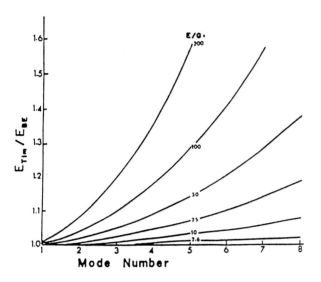

Fig. 4—Correction factors required to correct modulus values from resonant-frequency measurements using Bernoulli-Euler theory to values using Timoshenko theory. Factors are plotted as a function of mode number for several values of E/G and L/t = 100. From Ref. 23

volume element.[8] If this approach is to yield meaningful results, the scale of the inhomogeneity in the composite must be much smaller than the characteristic length associated with the structure and the characteristic wavelength associated with the dynamic-stress distribution. This is almost always the case with vibration testing, due to the relatively low frequencies involved. High-frequency wave propagation experiments will produce erroneous values of 'effective moduli' when the wavelength approaches the size of the inhomogeneity, however.

Single-Degree-of-Freedom Curve-Fitting Methods

As shown in any vibrations textbook, the parameters describing the vibration response of a single-degree-of-freedom (SDOF) spring-mass-damper system may be used in reporting damping-test results.[28,29] SDOF-damping parameters may be estimated by curve-fitting to the measured response of material specimens in either free vibration or forced vibration if a single mode can be isolated for the analysis. Approximate relationships between the loss factor from complex-modulus notation and these SDOF damping parameters exist for lightly damped systems;[30-32] such relationships will be used often in the following sections.

Free-Vibration Methods

Observations of the free-vibration response of a damped system are often used to characterize the damping in the system. If the specimen is released from some initial static displacement or if a steady-state forcing function is suddenly removed, the resulting free vibration response (Fig. 5) may be analyzed using the logarithmic decrement, an SDOF damping parameter. The logarithmic decrement is

$$\Delta = \frac{1}{n} \ln \frac{x_o}{x_n} \qquad (3)$$

where x_o and x_n are amplitudes measured n cycles apart. Equation (3) is based on the assumption of viscous damping, but for small damping, the loss factor from complex modulus notation may be approximated by[32]

$$\eta = \frac{\Delta}{\pi} \qquad (4)$$

Commonly used modes of testing include torsional pendulum oscillation[5,33] as shown in Fig. 6 and flexural vibration of beams or reeds.[34,35] Errors may result if more than one mode of vibration is significant in the free-vibration response, or if the data are taken at large amplitudes where air damping is present. The storage modulus is found by substituting the measured frequency of oscillation, specimen dimensions and density into the frequency equation for the particular specimen configuration and boundary conditions.[33,34]

Fig. 5—Free-vibration decay curve for logarithmic-decrement calculation

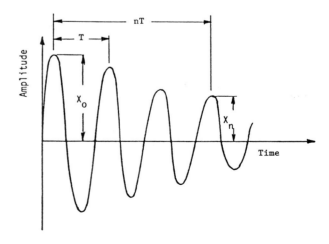

155

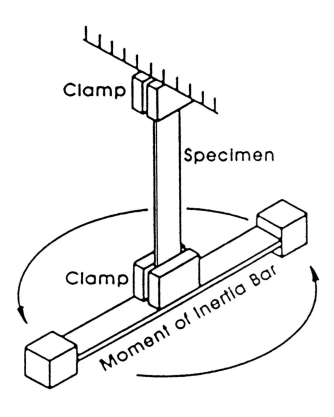

Fig. 6—Torsional-pendulum apparatus for free-vibration decay test. From Ref. 5

$$\eta \cong \frac{a}{b} \qquad (5)$$

and

$$E' \cong \frac{b}{c} \qquad (6)$$

where *a, b* and *c* are dimensions of the stress-strain ellipse in Fig. 7. Exact relationships exist for high-damping cases where the ellipses are not narrow.[30-32] Since the loss factor is the tangent of a small-phase angle, even small amounts of phase lag in the measurement system will cause errors. For example, electromechanical XY recorders may introduce phase-lag errors at frequencies above 1 Hz. Recorder phase lag can be checked easily by just plotting load vs. load—if the plot is not a straight 45-deg. line, the recorder is introducing its own phase lag.

Similar fixed frequency oscillation tests are the basis of several commercially available dynamic-mechanical analyzers which are referred to in the ASTM standards.[5] In these systems, data reduction is automated by interfacing a desktop computer with the measurement transducers. Some of these systems can also be used in the flexural and torsional modes. These systems were developed primarily for polymer testing, however, and their limitations for composite testing have already been discussed under 'Special Considerations.'

Fig. 7—Exaggerated elliptical hysteresis loop from fixed-frequency forced-oscillation test

Forced-Vibration Methods

Forced-vibration techniques are often more useful than free-vibration techniques when the control of amplitude and frequency is desired. Excitation may be sinusoidal, random or impulsive, and response may be analyzed in either the time domain or the frequency domain.

The simplest forced-vibration technique involves the measurement of uniaxial hysteresis loops during low-frequency sinusoidal oscillation of a tensile specimen in a servohydraulic mechanical-testing machine.[30,36,37] The elliptical hysteresis loops are just the Lissajous patterns formed by plotting the sinusoidally varying load (or stress) vs. the corresponding strain (Fig. 7). Not surprisingly, the complex-modulus notation also leads to the equation for an ellipse in the stress-strain plane.[31,32] For most composites, the loss factors are small enough that the ellipses are very narrow, and the components of the complex modulus can be approximated by the equations

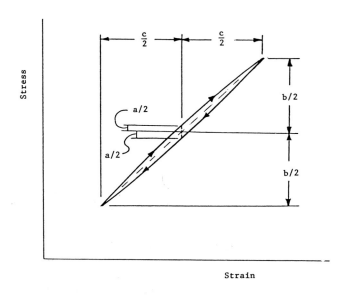

With the forced-vibration techniques discussed above, data are obtained at the frequency of oscillation of the exciter in the testing machine, which may or may not be a resonant frequency of the specimen. If the forcing frequency is tuned to a resonant frequency of the specimen, the relationship between the input and the response takes on a special form; this is the basis of the so-called resonant dwell method.[13,15,18,38,39] A resonant dwell apparatus for double-cantilever beams is shown in Fig. 8. In this case, when the specimen is excited at resonance, the loss factor is

$$\eta = C_n \, \frac{a(o)}{a(L)} \qquad (7)$$

where C_n is a constant for the nth mode, $a(o)$ is the base displacement amplitude, and $a(L)$ is the tip displacement amplitude.[38] Again, the storage modulus is found from the resonant frequency. A later modification involved the measurement of both force and acceleration at the driving point.[40]

By varying the forcing frequency, the so-called frequency response curve (or response spectrum) for the specimen can be swept out in the frequency domain, as shown in Fig. 9. The peaks in the curve represent resonant frequencies, and SDOF curve-fitting techniques such as the half-power bandwidth[28,29] can be used at these frequencies. The loss factor here is equal to

$$\eta = \frac{\Delta f}{f_n} = \frac{1}{Q} \qquad (8)$$

where Q is the quality factor (an electrical engineering term), Δf is the bandwidth at the half-power points on the resonant peak, and f_n is the peak frequency. Either the frequency-domain transfer function (the ratio of the response spectrum to the input spectrum) or the response spectrum alone can be used for this SDOF analysis.[28,29]

Digital frequency-spectrum analyzers based on the microcomputer-implemented Fast Fourier Transform (FFT) algorithm have made it possible to generate frequency-response curves in real time; techniques based on such analyzers will be discussed later. Curves such as those in Fig. 9 do not have sufficient frequency resolution for accurate determination of the half-power bandwidth; thus smaller frequency spans centered on the peak frequency are required. Most FFT analyzers have a band-selectable (or 'zoom') analysis feature that makes such high resolution possible.

The so-called 'swept-sine' test involves the use of variable-frequency sinusoidal excitation to sweep out the frequency-response curves.[6,32,34,41,42] Although this method is generally very slow, the input power is concentrated at one frequency and this may be necessary to move large specimens. Random[16,43] or impulsive[16,44] excitation is a much faster way to generate the frequency-response curve, but the excitation energy is broad-band in nature and it may be difficult to move large specimens. The previously discussed resonant dwell apparatus shown in Fig. 8 was also used with random excitation to generate the frequency-response curves.[16] Flexural and extensional versions of an impulse-frequency response apparatus based on a desktop computer interfaced with an FFT analyzer are shown in Figs. 10 and 11, respectively.[44] Flexural and extensional techniques were used to obtain the frequency dependence of the complex modulus of unidirectional composites along fiber directon as shown in Figs. 12 and 13, while the extensional technique was used to test the same composites at different fiber orientations as shown in Figs. 14 and 15.[45] A torsional version of this apparatus has also been developed.[46] Results from resonant dwell, random and impulse techniques show good agreement, as shown in Fig. 1.[16] The impulse technique has also been successfully used to characterize damage and degradation in composites and adhesive joints.[47-49]

All of the techniques described up to now have involved the determination of a single dynamic property from a single resonant frequency, but recent work has explored the problem of multiple property determination. For example, at least four resonant frequencies measured by the impulse technique have been used along with plate vibraton theories to determine the four independent elastic constants of orthotropic composite plates.[50,51] The portability of the equipment used with the impulse method makes it attractive for field testing of large structures as well.

Concluding Remarks

Numerous vibration-test techniques exist for dynamic-mechanical testing of composites and other structural materials. Factors such as the stress state, frequency range, material configuration, coupling between different modes of deformation, parasitic

Fig. 8—Double-cantilever beam specimen for resonant dwell test. From Ref. 38

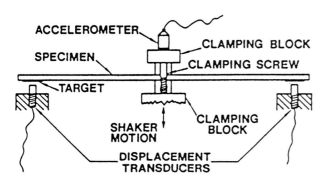

damping and the accuracy of analytical models used in data reduction must be considered before a method is selected for a particular application. Although standards exist for dynamic-mechanical testing of polymers, none exist at this time for composites. It is hoped that this article will provide guidance until standards can be developed.

Fig. 9—Typical specimen-transfer function vs. frequency, or frequency response curve. From Ref. 16

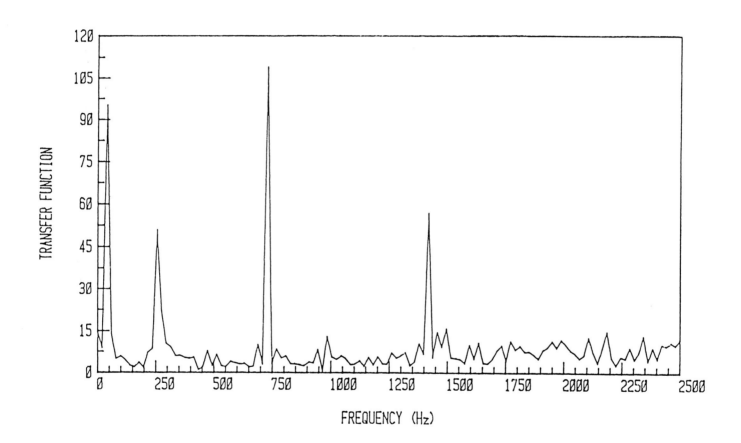

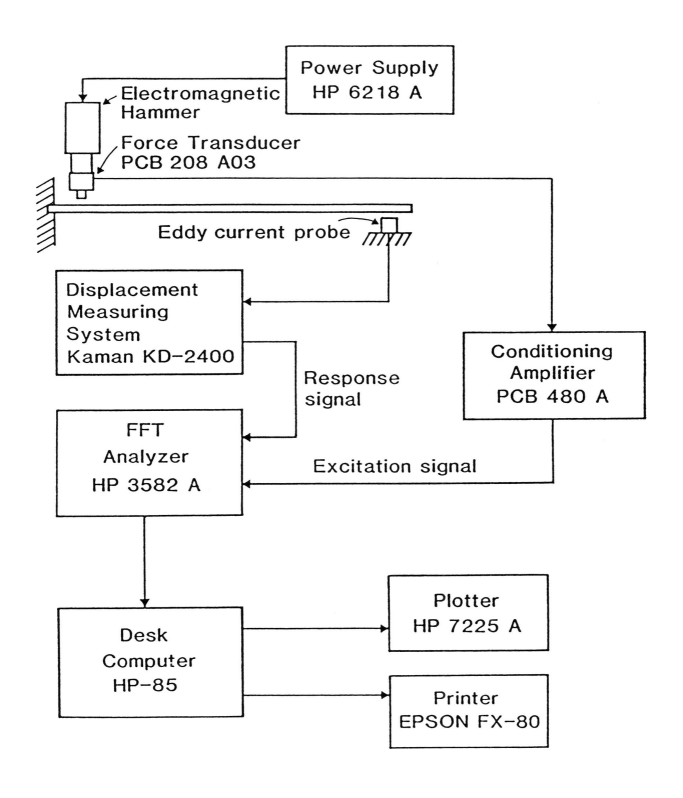

Fig. 10—Impulse-frequency response test apparatus for flexural vibration. From Ref. 44

Fig. 11—Impulse-frequency response test apparatus for extensional vibration. From Ref. 44

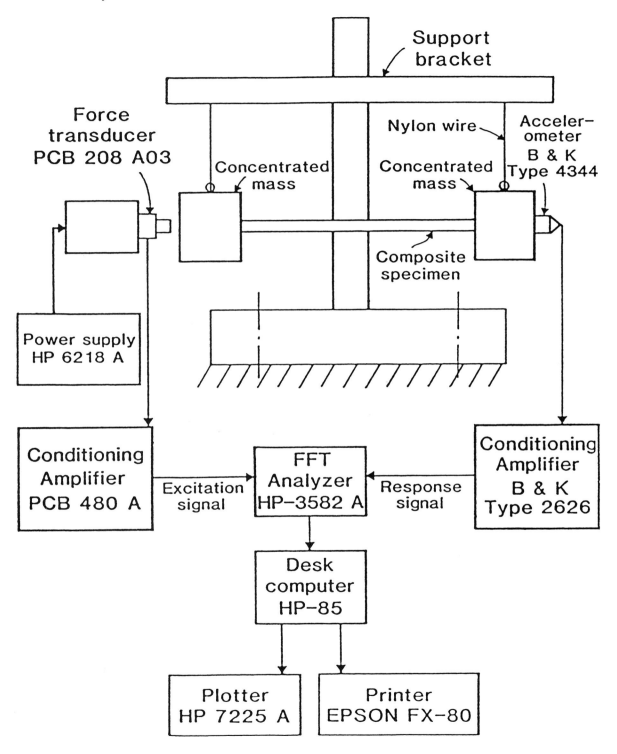

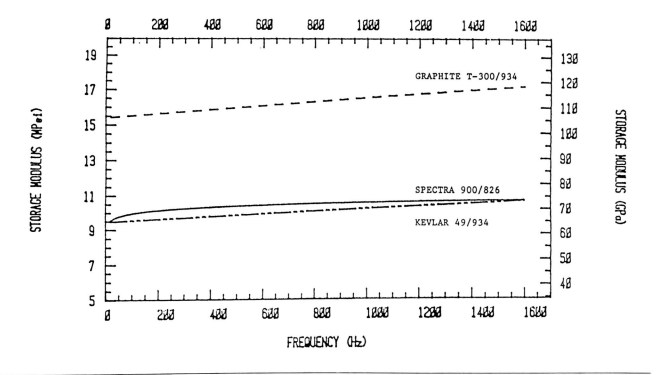

*Fig. 12—Comparison of storage modulus vs. frequency for three
unidirectional composites with 0-deg fiber orientation. From Ref.
45*

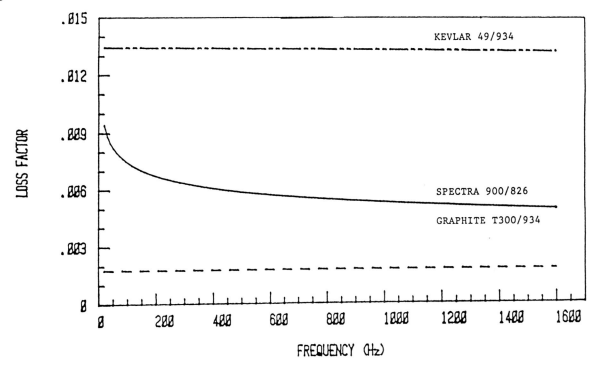

*Fig. 13—Comparison of loss factor vs. frequency for three
unidirectional composites with 0-degree fiber orientation. From
Ref. 45*

Fig. 14—*Comparison of storage modulus vs. fiber orientation for three unidirectional composites. From Ref. 45*

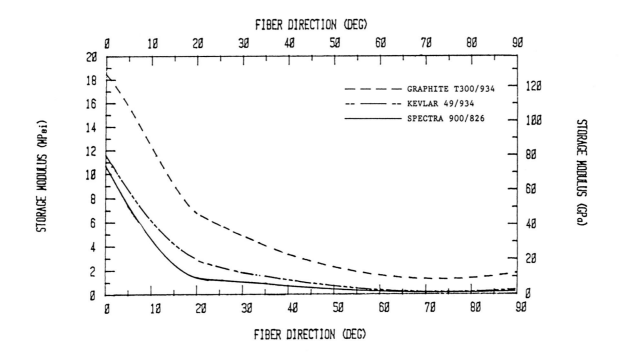

Fig. 15—*Comparison of loss factor vs. fiber orientation for three unidirectional composites. From Ref. 45*

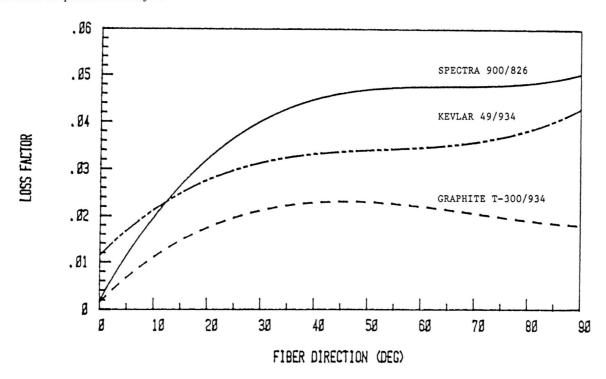

References

1. Gibson, R.F., "Dynamic Mechanical Properties of Advanced Composite Materials and Structures," The Shock and Vibration Digest, **19** (7), 13-22 (1987).

2. Bert, C.W. and Clary, R.R., "Evaluation of Experimental Methods for Determining Dynamic Stiffness and Damping of Composite Materials," Composite Materials: Testing and Design (Third Conf.), ASTM STP 546, ASTM, 250-265 (1974).

3. Bert, C.W., "Composite Materials: A Survey of the Damping Capacity of Fiber Reinforced Composites," Damping Applications for Vibration Control, ed. P.J. Torvik, ASME, AMD Vol. 38, 53-63 (1980).

4. Read, B.E. and Dean, G.D., The Determination of Dynamic Properties of Polymers and Composites, Adam Hilger, Ltd., Herts, England (1978).

5. Standard Practice for Determining and Reporting Dynamic Mechanical Properties of Plastics, ASTM Standard D 4065, ASTM (1982).

6. Standard Method for Measuring Vibration Damping Properties of Materials, ASTM Standard E 756, ASTM (1983).

7. Schapery, R.A., "Viscoelastic Behavior and Analysis of Composite Materials," Composite Materials: Vol. 2, Mechanics of Composite Materials, ed. G.P. Sendeckyj, Academic Press (1974).

8. Christensen, R.M., Mechanics of Composite Materials, John Wiley & Sons, New York (1979).

9. Jones, R.M., Mechanics of Composite Materials, McGraw-Hill, New York (1975).

10. Hasin, Z., "Complex Moduli of Viscoelastic Composites I: General Theory and Application to Particulate Composites" Int. J. Solids and Structures, **6**, 539-552 (1970).

11. Hashin, Z., "Complex Moduli of Viscoelastic Composites II", Int. J. Solids and Structures, **6**, 797-807 (1970).

12. Gibson, R.F., "Development of Damping Composite Materials," 1983 Advances in Aerospace Structures, Materials and Dynamics, AD-06, ASME, 89-95 (1983).

13. Gibson, R.F. and Plunkett, R., "A Forced Vibration Technique for Measurement of Material Damping," EXPERIMENTAL MECHANICS, **11**, (8), 297-302 (1977).

14. Baker, W.E., Woolam, W.E. and Young, D., "Air and Internal Damping of Thin Cantilever Beams," Int. J. Mechanical Engineering Science, **9**, 743-766 (1967).

15. Granick, N. and Stern, J.E., "Material Damping of Aluminum by a Resonant Dwell Technique," NASA TN D2893, (1965).

16. Suarez, S.A., Gibson, R.F. and Deobald, L.R., "Random and Impulse Techniques for Measurement of Damping in Composite Materials," EXPERIMENTAL TECHNIQUES, **8** (10), 19-24 (1984).

17. Gibson, R.F., Yau, A. and Riegner, D.A., "Vibration Characteristics of Automotive Composite Materials," Short Fiber Reinforced Composite Materials, ASTM STP 772, ASTM, 133-150 (1982).

18. Rogers, J.D. and McConnell, K.G., "Damping in Aluminum-Filled Epoxy Using Two Different Flexural Testing Techniques," Int. J. Analytical and Experimental Modal Anal., 8-17 (Oct. 1986).

19. Lee, J.M. and McConnell, K.G., "Experimental Cross-Verification of Damping in Three Metals," EXPERIMENTAL MECHANICS, **15** (9), 221-225 (1975).

20. Timoshenko, S.P., Young, D.H., and Weaver, W., Jr., Vibration Problems in Engineering, John Wiley & Sons, New York, NY (1974).

21. Huang, T.C. and Huang, C.C., "Free Vibrations of Viscoelastic Timoshenko Beams," J. Appl. Mech., **38**, Series E (2), 515-521 (1971).

22. Nakao, T., Okano, T. and Asano, I., "Theoretical and Experimental Analysis of Flexural Vibration of the Viscoelastic Timoshenko Beam," J. appl. Mech., **52** (3), 728-731 (1985).

23. Dudek, T.J., "Young's and Shear Moduli of Unidirectional Composites by a Resonant Beam Method," J. Composite Mat., **4**, 232-241 (1970).

24. Kalyanasundaram, S., Allen, D.H. and Schapery, R.A., "Dynamic Response of a Viscoelastic Timoshenko Beam," Proc. 28th AIAA/ASME/ASCE/AHS Structures, Structural Dynamics and Materials Conf., Monterey, CA, Paper No. AIAA-87-0890-CP (April 1987).

25. Horgan, C.O., "Some Remarks on Saint-Venant's Principle for Transversely Isotropic Composites," J. Elasticity, **2** (4), 335 (1972).

26. Choi, I. and Horgan, C.O., "Saint-Venant's Principle and End Effects in Anisotropic Elasticity," J. Appl. Mech., **44**, 424 (1977).

27. Horgan, C.O., "Saint-Venant End Effects in Composites,", J. Comp. Mat., **16**, 411-422 (1982).

28. Thomson, W.T., Theory of Vibration with Applications, Prentice-Hall, Englewood Cliffs, NJ (1972).

29. Meirovitch, L., Elements of Vibration Analysis, McGraw-Hill, New York, NY (1975).

30. Lazan, B.J., Damping of Materials and Members in Structural Mechanics, Pergamon Press, New York, NY (1968).

31. Nashif, A.D., Jones, D.I.G., and Henderson, J.P., Vibration Damping, John Wiley & Sons, New York, NY (1985).

32. Soovere, J. and Drake, M.L., Aerospace Structures Technology Damping Design Guide: Volume I - Technology Review, AFWAL-TR-84-3089 Vol. I, Air Force Wright Aeronautical Labs, Wright-Patterson AFB, OH (1985).

33. Ward, I.M., Mechanical Properties of Solid Polymers, John Wiley & Sons, New York, NY, 2nd Ed. (1983).

34. Schultz, A.B. and Tsai, S.W., "Dynamic Moduli and Damping Ratios in Fiber-Reinforced Composites," J. Comp. Mat., **2** (3), 368-379 (1968).

35. Crawley, E.F. and Mohr, D.G., "Experimental Measurements of Material Damping in Free Fall with Tunable Excitation," AIAA Journal, **23** (1), 125-131 (1985).

36. Gibson, R.F., "Vibration Damping Characteristics of Graphite/Epoxy Composites for Large Space Structures," Proc. 3rd Large Space Systems Tech. Rev., NASA Conf. Publ. 2215, Part 1, 123-132 (1982).

37. Ray, A., Kinra, V., Rawal, S. and Misra, M., "Measurement of Damping in Continuous Fiber Metal Matrix Composites," Role of Interfaces on Material Damping, ASM, 95-102 (1985).

38. Gibson, R.F., Yau, A. and Riegner, D.A., "An Improved Forced Vibration Technique for Measurement of Material Damping," EXPERIMENTAL TECHNIQUES, **6**, 10-14 (1982).

39. Rogers, J.D., Zachary, L.W. and McConnell, K.G., "Damping Characterization of a Filled Epoxy Used for Dynamic Structural Modeling," EXPERIMENTAL MECHANICS, **26** (3), 283-291 (1986).

40. Rogers, J.D. and McConnell, K.G., "Instrumentation for Determination of Material Damping from Driving Point Measurements," Role of Interfaces on Material Damping, ASM, 103-110 (1985).

41. Adams, R.D. and Bacon, D.G.C., "The Dynamic Properties of Unidirectional Fibre Reinforced Composites in Flexure and Torsion," J. Comp. Mat., **7**, 53-67 (1973).

42. Adams, R.D. and Bacon, D.G.C., "Measurement of the Flexural Damping Capacity and Dynamic Young's Modulus of Metals and Reinforced Plastics," J. Physics D: Appl. Phys., **6**, 27 (1973).

43. Soovere, J., "Dynamic Response of Flat Integrally Stiffened Graphite/Epoxy Panels Under Combined Acoustic and Shear Loads," Recent Advances in Composites in the U.S. and Japan, ASTM STP 864, ASTM, 281-296 (1985).

44. Suarez, S.A. and Gibson, R.F., "Improved Impulse-Frequency Response Techniques for Measurement of Dynamic Mechanical Properties of Composite Materials," J. Testing and Eval. **15** (2), 114-121 (1987).

45. Gibson, R.F., Rao, V.S. and Mantena, P.R., "Vibration Damping Characteristics of Highly Oriented Polyethylene Fiber Reinforced Epoxy Composites," Advanced Materials Technology '87, Proc. 32nd Int. SAMPE Symp., 231-244 (1987).

46. Place, T.A., Gibson, R.F., Mantena, P.R., Perez, N. and Bobeck, G., "The Influence of Thermal Degradation on Internal Damping of a Ni-Mo-Fe-B Rapidly Solidified Alloy," 116th Annual TMS-AIME Mtg., Denver, CO (Feb. 1987).

47. Mantena, P.R., Place, T.A. and Gibson, R.F., "Characterization of Matrix Cracking in Composite Laminates by the Use of Damping Capacity Measurements," Role of Interfaces on Material Damping, ASM, 79-94 (1985).

48. Mantena, P.R., Gibson, R.F., Place, T.A., Srivatsan, T.S. and Sudarshan, T.S., "Debond and Failure Characristics of Double Lap Adhesively Bonded Joint," Proc. 13th Annual Int. Symp. for Testing and Failure Analysis—ISTFA/87, Los Angeles, CA, ASM Int., 225-233 (Nov. 1987).

49. Mantena, P.R., Gibson, R.F. and Place, T.A., "Damping Capacity Measurements of Degradation in Advanced Materials," SAMPE Quarterly, 17 (3), 20-31 (1986).

50. DeWilde, W.P., Sol, H. and Van Overmeire, M., "Coupling of Lagrange Interpolation, Modal Analysis and Sensitivity Analysis in the Determination of Anisotropic Plate Rigidities," Proc. 4th Int. Modal Anal. Conf., Los Angeles, CA, 1058-1063 (Feb. 1986).

51. Deobald, L.R. and Gibson, R.F., "Determination of Elastic Constants of Orthotropic Plates by a Modal Analysis/Rayleigh Ritz Technique," J. Sound and Vibraton, 124 (2), 269-283 (1988).

Section VIIF

Edge Replication for Laminated Composites

by Alton L. Highsmith

Introduction

When studying failure processes in laminated composites, it is useful to monitor the progression of damage throughout the load history of the material. A variety of test techniques can provide information about damage accumulation, but many of these techniques infer the damage states from some sophisticated interpretation of the raw data. Only two techniques, X-ray radiography and edge replication, give a direct indication of the damage state. The present chapter provides a brief description of edge replication, a technique for the *in situ* documentation of damage in laminated composites.

Edge replication was first applied to composites by Stalnaker and Stinchcomb[1] and Masters and Reifsnider[2]. The procedure itself is relatively simple. Cellulose acetate is softened with acetone and then pressed against the edge of the specimen. The specimen edge acts as a mold to which the cellulose acetate conforms. When the acetone evaporates, the cellulose acetate hardens, and can be pulled away from the specimen. The hardened cellulose acetate provides a permanent impression, or replica, of the specimen edge. Once obtained, a replica may be studied at leisure in order to assess the damage state in the laminate at the point in the load history when the replica was made. Only a brief interruption of a quasi-static or fatigue test is required for the actual replication process.

As with many experimental techniques, there is a degree of artistry involved in making an edge replica. A certain 'touch' is acquired with experience. A particular procedure for making edge replicas of straight-sided coupon type specimens is described below. This procedure is properly considered a guideline, as any number of variations on the procedure may yield good edge replicas. However, the given procedure has proven successful for the novice with a minimum of frustration.

Specimen Preparation

Figure 1(a) is a photomicrograph of an edge replica taken from a $[0/\pm45/90]_s$ graphite/epoxy laminate in the as-received condition. The specimen was cut from a panel using a water-cooled diamond wheel

ficult to distinguish the 45, -45, and 90-degree plies. Also, some rather large diagonal scratches are visible. These scratches are a result of slight wobbling of the saw blade during the cutting process. Polishing the specimen edge results in considerable improvement of the image quality of the replica.

Fig. 1—Edge replicas from a quasi-isotropic graphite/epoxy specimen (a) before and

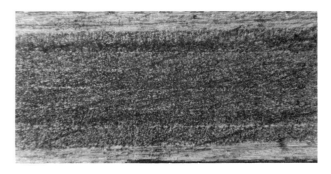

(b) after polishing.

Figure 1(b) is a photomicrograph of an edge replica taken from the same quasi-isotropic specimen after polishing the specimen edge. To polish the edge, an abrasive is stretched over a flat glass plate. The specimen edge is stroked across the abrasive. Care is required to keep the specimen normal to the glass plate so that bevelling of the specimen edge is avoided. The specimen is polished with successively finer abrasives. The specimen of Fig. 1(b) was wet sanded on 320, 400 and 600 grit silicon-carbide wet/dry papers, and then polished on a felt polishing cloth soaked with a slurry of 5-micron alumina powder and water. All of the plies in the figure are distinct, and fiber ends in the off-axis plies are visible. The resin rich fiber-matrix interfaces between plies can also be seen.

Initially, it is necessary to monitor progress during the polishing process to insure that the edge has been polished sufficiently at each level of abrasive. This is typically done by making edge replicas throughout the process and examining them to make sure that any given abrasive is sufficient to remove existing scratches. If scratches persist when using one abrasive, a coarser abrasive is required. With experience, the polishing procedure can be standardized for a given material system. For example, the graphite/epoxy specimen described above was polished for 50 strokes on each abrasive. A similar procedure will provide good results for any laminate of the same material system, provided the as-received edges are in comparable conditions. Also, it is possible to polish the edge too much. The two phases in

the composite polish at different rates. This is why the fiber-ends and interfaces are visible. However, if even finer aluminum powder is used, the specimen surface can be made essentially flat, and these details will not be visible. Generally, it is useful to be able to distinguish fiber ends and interfaces.

The Replication Process

To make an edge replica, a piece of cellulose acetate is attached to the specimen edge using adhesive tape. The cellulose acetate is a thin film on the order of 3-10 mils thick. It is available in a sheet form and also in a roll form referred to as replicating tape, and can be purchased from Ernest F. Fullam, Inc. Typically the gage section of the specimen is marked on its surface so that the acetate can be positioned appropriately. An additional mark in the center of the gage section can be used as a guide for marking the replica itself, so that a particular location can be observed during microscopic examination of a sequence of replicas.

Next, a syringe is filled with acetone. The acetone is injected between the specimen edge and the cellulose acetate. Figure 2 shows a piece of cellulose acetate attached to a specimen edge. A syringe is shown in position to inject the acetone. Injecting the acetone is a delicate operation, as only a small

Fig. 2—Specimen and replicating tape prepared for the injection of acetone.

Fig. 3—An eraser presses the replica against the specimen edge while the replica hardens.

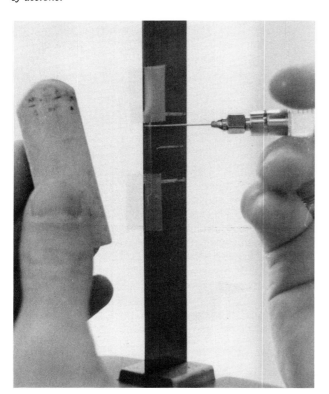

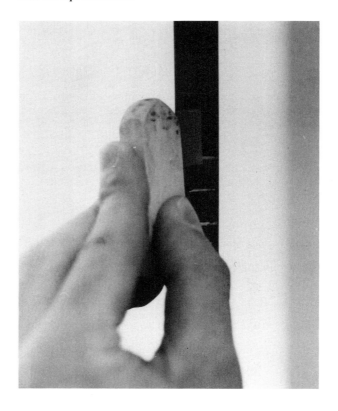

amount of acetone is needed to make the replica. Using too much acetone effectively 'washes out' the surface details in the cellulose acetate. Glass syringes tend to operate more smoothly than plastic ones, and thus provide better control of the flow of acetone.

After injecting the acetone, the cellulose acetate must be pressed against the specimen edge. The most consistent results are obtained by using a flat compliant object to distribute the pressure evenly across the replicating tape. Figure 3 shows an eraser being used to apply pressure to the replicating tape. Notice that the eraser is also seen in Fig. 2. It is important to apply pressure to the cellulose acetate immediately after the acetone is injected but it is not as urgent as is implied by Fig. 2. There is sufficient time to pick up the eraser and press it against the specimen, if such a procedure is more comfortable. Pressure should be maintained for about 30 seconds. The cellulose acetate should then be allowed to harden for about two minutes before the replica is removed for the specimen. Replicas are best kept taped to a glass slide. They should be held flat, as they tend to curl with time.

One additional consideration is how the specimen should be loaded during the replication process itself. The photomicrograph in Fig. 4(a) is for a replica of a [0/±45/90]$_s$ graphite/epoxy specimen that had been subjected to a quasi-static load of 60 ksi. The replica does provide indications of matrix cracking in the 90-and -45 deg plies, as indicated by the sharp dark lines that traverse the thickness of these plies. The cracks in the 90-deg plies are essentially perpendicular to the 0-deg direction, while the cracks in the -45-deg plies tend to be oriented at about a 45-deg angle to the 0-deg direction. This orientation in the -45-deg plies is due to the free-edge effect. Notice that the cracks are clearly visible only in the 90-deg ply and the upper -45-deg ply. Figure 4(b) shows the same region in a replica taken while the specimen was subjected to a 10-ksi load. This modest load opens the cracks sufficiently that cracks can now be seen in both -45-deg plies. Those cracks that were visible in Fig. 4(a) are more distinct in Fig. 4(b). Clearly, the applied load has enhanced the details in the replica. Note also that a relatively small load can provide significant improvement. This is especially important when testing specimens near their ultimate tensile strength, where maintaining the peak load might prove hazardous to the experimenter.

There are a number of other factors which influence replica quality. Some involve skill in making replicas. For example, the proper amount of acetone must be injected, and the cellulose acetate must be pressed against the specimen without sliding along the specimen. These factors can be controlled through practice. It is also possible for damage to influence replica quality. In particular, large edge delaminations tend to hinder the replication process. These damage events disturb the flow of the acetone,

and the resulting replicas reflect the disturbed flow field. Features resembling wakes appear in the replicas. Such features are not a result of poor replication technique.

Fig. 4—Replicas taken from a damaged quasi-isotropic specimen while (a) the specimen was unloaded and (b) the specimen was subjected to a 10 Ksi. load.

(a)

(b)

Viewing and Displaying Replicas

There are several techniques that can be used to view and display edge replicas. For the most part, these techniques depend on the optical properties of edge replicas. When light is projected through an edge replica, less light passes through the thicker parting of the cellulose acetate. For example, during the replication process, the softened cellulose acetate flows into any ply cracks present in the specimen. When the replica is removed from the specimen, there are bulges or 'high spots' in the replica corresponding to those ply cracks. When light is projected through the replica, the cracks are indicated by dark lines. Probably the simplest method for displaying replicas is to use the replica itself as a photographic negative. The replica is placed in a photographic enlarger and printed. Figure 5(a) is a photographic print made directly from a replica. There is a 'color reversal' of the printing process, so that the matrix cracks appear as white lines and the bulk of the replica is dark. While this technique is simple, it provides relatively low resolution. Further, considerable magnification of the replica is required in the enlarging process.

A replica can also be viewed using a microscope. The best results are obtained by using transmitted light instead of reflected light. One advantage of this system is that the large magnifications required for a detailed inspection of a replica are readily available. If the microscope is equipped with a traversing mechanism, it is a simple matter to scan the length of the replica. Standard microscope mounts are

available for 35mm cameras so that it is a straightforward matter to photograph portions of the magnified image. The photograph in Fig. 5(b) was taken using a microscope and 35mm camera. This image shows considerably more detail than that of Fig. 5(a), obtained by printing the replica directly.

One of the most significant contributions to edge replication technology was provided by Kriz[3], who first suggested viewing replicas with a microfiche reader. Many microfiche readers are equipped with 50 X lenses which are well suited for viewing edge replicas. They also have a traversing mechanism which allows scanning the length of the replica. In short, the microfiche reader provides excellent image quality with a minimum of effort and eye strain. A small portable unit is especially useful for inspection of the replicas immediately after they are made. Replica quality can be quickly checked before mechanical testing is continued. In addition, the more sophisticated microfiche readers also have provisions for making photocopies of the image. Such devices can be found in the microtext section of a library. Figure 5(c) was produced on a microfiche reader with photocopying capabilities. While the image does not have as much detail as one obtained using a microscope, it is more than adequate for observing ply cracks.

Applications

The edge replication technique can be applied to a wide variety of materials. The replicas presented here were obtained from grapite/epoxy specimens. Other researchers have applied the technique to material systems ranging from cord rubber[4] to glass/epoxy[5]. The applications to glass/epoxy is especially important since these materials are not well suited to X-ray radiography. This is because the glass fibers are rather opaque to X-rays. Thus, edge replication may be the only way to nondestructively monitor damage development directly.

The technique has also been used to study fiber-matrix debonding in silicon carbine/titanium metal matrix composites[6]. The edge replicas contain sufficiently detailed information to indicate whether or not the fiber and matrix have separated. However, such a high-resolution examination requires the use of a scanning-electron microscope. The procedure is similar to conventional microscopy, except that the replicas must be coated with a conductive material (typically gold) before they can be viewed.

Damage in the interior of a specimen can be studied by sectioning the specimen and replicating the sectioned edge. Clearly this is a destructive process and can only be done once for a given specimen. Also, the user must be aware that some damage may be introduced during the sectioning process itself.

Fig. 5—Images of a replica obtained from (a) direct printing using the replica as a negative, (b) photomicroscopy, and (c) a photocopying microfiche reader.

(a)

(b)

(c)

Summary

Edge replication is a technique for documenting damage development in composite materials. The technique provides a direct indication of damage present at the specimen edge. It requires no specialized equipment or supplies other than the cellulose-acetate film used to make the replica. Because of its simplicity and applicability to a wide range of material systems, edge replication is an ideal 'standard experimental method' for the composites-testing laboratory.

References

1. Stalnaker, D.O. and Stinchcomb, W.W., "Load History-Edge Damage Studies in Two Quasi-Isotropic Graphite Epoxy Laminates," Composite Materials: Testing and Design (Fifth Conference), ASTM STP 674, ed. S.W. Tsai, ASTM, 620-641 (1979).

2. Reifsnider, K.L. and Masters, J.E., "Investigation of Characteristic Damage States in Composite Laminates," ASME Paper No. 78-WA/Aero-4, Winter Ann. Mtg. (1978).

3. Kriz, R.D. and Stinchcomb, W.W., "Effects of Moisture, Residual Thermal Curing Stresses, and Mechanical Load on the Damage Development in Quasi-Isotropic Laminates," Damage in Composite Materials, ASTM STP 775, ed. K.L. Reifsnider, ASTM, 63-80 (1982).

4. Stalnaker, D.O., Kennedy, R.H., and Ford, J.L., "Interlaminar Shear Strain in a Two-Ply Balanced Cord Rubber Composite," EXPERIMENTAL MECHANICS, 20 (3), 87-94 (1980).

5. Highsmith, A.L. and Reifsnider, K.L., "Stiffness-Reduction Mechanisms in Composite Laminates," Damage in Composite Materials, ASTM STP 775, ed. K.L. Reifsnider, ASTM, 103-117 (1982).

6. Johnson, W.S., Lubowinski, S.J., Highsmith, A.L., Brewer, W.D., and Hoogstraten, C.A., "Mechanical Characterization of SCS-6/Ti-15-3 Metal Matrix Composites at Room Temperature," NASA Tech. Memorandum NASP TM-1014 (April 1988).